AQUIFER TEST MODELING

AQUIFER TEST MODELING

William C. Walton

CRC Press is an imprint of the
Taylor & Francis Group, an informa business

CRC Press
Taylor & Francis Group
6000 Broken Sound Parkway NW, Suite 300
Boca Raton, FL 33487-2742

First issued in paperback 2019

ISBN-13: 978-1-4200-4292-4 (hbk)
ISBN-13: 978-0-367-38991-8 (pbk)

Library of Congress Cataloging-in-Publication Data

Walton, William Clarence.
Aquifer test modeling / William C. Walton.
p. cm.
Includes bibliographical references and index.
ISBN 1-4200-4929-0 (alk. paper)
1. Aquifers--Testing--Mathematical models. 2. Groundwater flow--Measurement--Mathematics. I. Title.

GB1199.W345 2006
628.1'14015118--dc22 2006050487

Visit the Taylor & Francis Web site at
http://www.taylorandfrancis.com

and the CRC Press Web site at
http://www.crcpress.com

To Ellen

Preface

The history and possible future of aquifer test analysis (modeling) are summarized by Renard (2005). Based on this summary, it would seem that mature traditional methods involving analytical integral and empirical equations such as those developed by Theis (1935) and Hantush and Jacob (1955) and type curve and straight line matching techniques for simple aquifer-well conditions are and will continue to flourish. However, more recent methods involving computer numerical Laplace inversion (Stehfest, 1970a, 1970b) of analytical equations for complicated aquifer-well conditions (see Moench and Ogata, 1984), improved conceptual model definition, and automatic regression calibration (see Doherty, 1994) are slowly finding acceptance and application in the groundwater industry. There seems to be a slowly growing trend toward the use of numerical models for analyzing aquifer test data (see Prince and Schneider, 1989).

In keeping with the summary, the purpose of this book is to further the application of numerical Laplace inversion analytical equations and numerical models through the use of public domain U.S. Geological Survey software programs WTAQ (Barlow and Moench, 1999) and MODFLOW (McDonald and Harbaugh, 1988), together with the public domain automatic parameter estimation software program PEST (Doherty, 2004). Toward that end, a protocol for organizing and simplifying conceptual model definition and data analysis is presented herein.

About the Author

Bill Walton has been active in the field of groundwater aquifer tests for the past 56 years. The first pumping test conducted and analyzed by Bill with the Theis equation (model) was at Eau Claire, Wisconsin in the fall of 1949. Since then, he has participated in aquifer tests at numerous sites throughout Canada and the United States and in Haiti, El Salvador, Dominican Republic, Libya, and Saudi Arabia. Bill has firsthand experience with aquifer tests in unconsolidated deposits, limestones, dolomites, and sandstones, as well as volcanic, igneous, and metamorphic rocks.

Bill graduated in 1948 with a B.S. in civil engineering from Lawrence Technological University at Southfield, Michigan. He also attended Indiana University, University of Wisconsin, Ohio State University, and Boise State University majoring in geology.

Bill's employment includes 2 years as a water well contractor in Detroit, Michigan; 1 year as a hydraulic engineer with the U.S. Bureau of Reclamation at Cody, Wyoming; 8 years as a hydraulic engineer with the U.S. Geological Survey at Madison, Wisconsin, Columbus, Ohio, and Boise, Idaho; 6 years as engineer in charge of groundwater research with the Illinois State Water Survey at Urbana, Illinois; 6 years as a consulting groundwater hydrogeologist with consulting firms including Shaefer and Walton in Columbus, Ohio; Camp Dresser and McKee, Inc. in Champaign, Illinois; and Geraghty and Miller in Champaign, Illinois; 3 years as a self-employed consultant in water resources in Mahomet, Illinois; 10 years as a professor of geology and geophysics and director of the Water Resources Research Center, University of Minnesota at Minneapolis, Minnesota; 2 years as water resources planning director, Minnesota State Planning Agency in St. Paul, Minnesota; and 5 years as executive director of the Upper Mississippi River Basin Commission in Minneapolis, Minnesota.

Bill served as a visiting scientist for the American Geophysical Union and the American Geological Institute and lectured at many universities throughout the United States. He also lectured at short courses sponsored for several years by the International Ground Water Modeling Center and presented several papers at professional society meetings in Europe. Bill has served as Minnesota's member of the Great Lakes Basin Commission, Souris-Red-Rainy River Basins Commission, Upper Mississippi River Basin Coordinating Committee, and Missouri Basin Inter-Agency Committee. He was also an advisor to the Land and Water Resources Committee of the Minnesota House of Representatives, a member of the Technical Advisory Committee of the Minnesota–Wisconsin Boundary Area Commission, member of the Citizen Advisory Committee of the Minnesota Environmental Quality Council, vice president of the National Ground Water Association and founding editor of the journal *Ground Water*, chairman of the Ground Water Committee of the Hydraulics Division of the American Society of Civil Engineers, member of the U.S. Geological Survey Advisory Committee on Water Data for Public Use, Consultant to the Office of Science and Technology at Washington, D.C., member of the Steering Committee of the International Ground Water Modeling Center, member of the Committee for the International Hydrological Decade, and an advisor to the U.S. delegation to the Coordinating Council of the International Hydrological Decade of UNESCO at Paris, France.

Bill is author of over 75 technical papers and 9 books including *Groundwater Resource Evaluation* (McGraw-Hill); *The World of Water* (Weidenfeld and Nicolson); *Practical Aspects of Groundwater Modeling* (National Ground Water Association); and *Groundwater Pumping Tests, Analytical Groundwater Modeling, Numerical Groundwater Modeling, Principles of Groundwater Engineering, Groundwater Modeling Utilities, Designing Groundwater Models with Windows Software,* and *Aquifer Test Analysis with Windows Software* (Lewis Publishers).

Table of Contents

1 Overview

Aquifer test modeling is herein defined as the process of searching for an aquifer test domain conceptual model that closely reproduces the measured response of an aquifer to a controlled discharge or slug change in the head. There are five major steps in the aquifer test modeling process:

1. Conceptual model definition
2. Modeling equation and software selection
3. Data adjustment
4. Data analysis
5. Model evaluation

A *conceptual model* is a space and time representation (approximation) of the groundwater flow system within an aquifer test domain that captures the essence of the groundwater system. An aquifer test domain is a volume of the groundwater flow system surrounding and within the influence of a pumped or slugged well. The aquifer test domain height is the groundwater flow system thickness, which can be 100 ft or more. A pumping test domain radius of influence can be 500 ft or more under unconfined aquifer conditions and 1000 ft or more under confined nonleaky, leaky, or fissure and block aquifer conditions. The slug test domain height can be 20 ft or more. A slug test domain radius of influence can be 20 ft or more.

A conceptual model organizes available groundwater system information and is based on the following:

- Hydrostratigraphic framework of the groundwater system (type, thickness, extent, heterogeneities, and boundaries of hydrostratigraphic units)
- Hydraulic characteristics of hydrostratigraphic units (hydraulic conductivity and storage)
- Pumped or slugged and observation well or piezometer characteristics (such as casing radius, effective radius, pump pipe radius, well-bore storage, and well-bore skin)
- Time dimensions (aquifer test elapsed time in analytical groundwater flow models or the number and lengths of stress periods and time steps in numerical groundwater flow models)
- Groundwater and surface water budgets (pump discharge or slug initial displacement and any recharge)
- Flow system temporal changes (such as water level fluctuations due to atmospheric pressure, stream stage changes, or tidal fluctuations)

Conceptual model definition should take into consideration the simplifying assumptions upon which the aquifer test mathematical modeling equations selected for use are based, the data-input requirements (file or interactive data entry) of the aquifer test software program selected for use, and the calibration (interactive or automatic parameter estimation) data-input requirements of the aquifer test software program selected for use.

The possible range of conceptual model layer thickness and extent as well as the most probable conceptual model layer thickness and areal extent are ascertained by carefully studying aquifer test well logs, cross sections, plan view maps, and groundwater level graphs. The possible range of conceptual model hydraulic parameter (horizontal and vertical hydraulic conductivity, storability, and specific yield) values, as well as the most probable conceptual model hydraulic parameter values, are selected based on hydrogeologic data, laboratory test data, logic, judgment, and information in published tables.

The most probable extent of the cone of influence around the pumped or slugged well, the duration of any well-bore storage impacts, the most probable extent of any well partial penetration impacts, and the most probable duration of delayed gravity drainage under unconfined conditions should be considered in selecting the possible range of conceptual models as well as the most probable conceptual model.

Modeling equation and software selection involves a review of available aquifer test mathematical modeling equations and associated aquifer and well conditions. There are two types of aquifer test mathematical modeling equations:

1. Analytical
2. Numerical

Analytical mathematical modeling equations are usually selected when the conceptual model consists of two or less layers and aquifer parameters are fairly uniform in space. Analytical mathematical modeling equations are also selected to guide the use of numerical modeling equations. Numerical mathematical modeling equations are usually selected when the conceptual model consists of more than two layers or the aquifer parameters are highly heterogeneous.

There are two types of analytical mathematical modeling equations:

1. Stehfest algorithm
2. Integral and empirical

Stehfest algorithm mathematical modeling equations can be applied with little difficulty to the analysis of both simple and more complex aquifer and pumping or slug conditions. Integral and empirical mathematical modeling equations are special cases of Stehfest algorithm mathematical modeling equations with certain parameter values assumed to be negligible. Integral and empirical mathematical modeling equations can be applied without difficulty to the analysis of data for simple aquifer and pumping or slug conditions.

If well-bore storage is appreciable in the conceptual model, dimensionless time-drawdown values calculated with integral modeling equations should be adjusted for well-bore storage. Otherwise, the ranges of pumping test data that are not likely to be affected appreciably by well-bore storage should be subjectively selected (filtered) for analysis. If the conceptual model has partially penetrating wells, dimensionless time-drawdown or normalized head values should be adjusted for the effects of partially penetrating wells. Otherwise, aquifer test data that are not likely to be affected appreciably by well partial penetration should be subjectively selected for analysis.

Both pumping test integral and Stehfest algorithm mathematical modeling equations assume that nonlinear well losses in the pumped well are negligible. If there are well losses, drawdowns in the pumped well should be adjusted for the effects of well loss. Both pumping test integral and Stehfest algorithm mathematical modeling equations assume the aquifer is infinite in areal extent within the radius of influence of the pumped or slugged well (aquifer test domain). If the aquifer is finite in extent in the conceptual model, dimensionless time-drawdown or normalized head should be adjusted for boundary effects. Otherwise, ranges of data not likely to be affected by boundaries should be subjectively selected for analysis.

Well-bore storage, well-bore skin, and partially penetrating wells are fully covered in Stehfest algorithm mathematical modeling equations. The use of Stehfest algorithm mathematical modeling equations in aquifer test analysis eliminates the need for any well-bore storage and well partial penetration adjustments and is highly recommended. Pumping test Stehfest algorithm mathematical modeling equations can be applied to slug tests by recognizing that dimensionless time-normalized head values are the first derivatives (slopes) of dimensionless time-drawdown values multiplied by a factor.

Most pumping test integral and Stehfest algorithm mathematical modeling equations assume the aquifer is horizontally isotropic. Horizontal anisotropic aquifer parameter values can be estimated by postprocessing pumping test analysis results based on horizontal isotropic assumptions.

Aquifer test numerical mathematical modeling equations commonly utilize the finite-difference approximation method. The conceptual model consists of a discretized grid of nodes and associated finite-difference cells (blocks) simulating one or more aquifer layers. Delayed gravity drainage under unconfined aquifer conditions can be simulated with 10 or more confined layers and 1 unconfined layer. Parameter values are assigned to grid cells and boundary conditions are simulated along or within grid cell borders.

Aquifer test time and pumping rates are discretized into small blocks of variable lengths and strengths to simulate the effects of well-bore storage. Time-drawdown values calculated with numerical models are calibrated against measured time-drawdown values. Calculated time-drawdown values can be converted to dimensionless time-drawdown values for use in type curve matching by using average aquifer hydraulic parameter values.

A large variety of public domain and commercial aquifer test analysis software is available with a broad range of sophistication. Primary software usually contains code to read input data files, code (calculation engine) to calculate either or both dimensionless or dimensional time-drawdown or time-normalized head values based on file input data, and code to generate output files for use with graphs or external word processor, spreadsheet, database, and automatic parameter estimation software. Primary software usually is distributed by governmental agencies. Sophisticated analytical and numerical aquifer test analysis software contains code for interactive computer screen input (preprocessor), a calculation engine, internal automatic parameter estimation code, and code to display calculation results on the computer screen or with a printer (postprocessor). Sophisticated software can also contain integrated word processor, spreadsheet, database, and graphics capabilities for seamless analysis. Sophisticated software is usually distributed commercially.

The U.S. Geological Survey distributes the fully documented primary analytical software WTAQ (Barlow and Moench, 1999) written in Fortran and containing state-of-the-art code for calculating analytical Stehfest algorithm mathematical model dimensionless or dimensional time-drawdown values with confined nonleaky or unconfined (water table) aquifer conditions. WTAQ source code (water.usgs.gov/nrp/gwsoftware/) can be expanded to cover confined leaky and confined fissure and block aquifer conditions using equations presented by Moench (1984). WTAQ source code can also be expanded to cover finite aquifer conditions using the image well theory (Ferris et al., 1962) and slug tests using the relationship between pumping and slug test responses (Peres et al., 1989).

The U.S. Geological Survey also distributes the fully documented primary numerical international standard software MODFLOW (McDonald and Harbaugh, 1988). Several versions of MODFLOW are available at water.usgs.gov/nrp/gwsoftware/. The latest version of MODFLOW-2000, written in Fortran, internally supports both interactive and automatic parameter estimation. The use of MODFLOW in conjunction with a U.S. Geological Survey radial-flow preprocessor (Reilly and Harbaugh, 1993a) to verify the results of analytical pumping test analysis is described in WTAQ documentation. PEST software and documentation are available at www.sspa.com.

Data adjustment consists of the review of the mathematical modeling equation assumptions and the adjustment of data for any departures (herein called external influences) from the assumptions. Erroneous conclusions about aquifer parameter values and boundaries can be reached if the impacts of any external influences are not removed before aquifer test data are analyzed with mathematical modeling equations. External influence fluctuations include those caused by groundwater flow through the aquifer test domain prior to the test (antecedent trend), atmospheric pressure changes, surface water (tidal, lake, or stream) stage changes, earth tides, earthquakes, applications of heavy loads (railroad trains or trucks), evapotranspiration, recharge from rainfall, and nearby pumped well pumping rate changes.

Data measurements prior to, during, and after the pumping or slug test are required for external influence adjustment. External influence data adjustments are usually based on data reference baselines that extend horizontally through the time immediately prior to test initiation and vertically through prominent external influence peaks and troughs. External influence data adjustment involves interpolation and extrapolation, which can be performed with data fitting computer programs in which a line or curve is fitted to data from past times and extended to estimate data for future times with linear or curvilinear polynomial approximation (regression) equations.

Data analysis involves conceptual models, mathematical modeling equations, one or more aquifer test formats and techniques, well function or drawdown-head calculation, and calibration. There are two analysis formats:

1. Dimensionless
2. Dimensional

Dimensionless format refers to the interactive calibration of measured time-drawdown or time-normalized head graphs to calculated dimensionless time-drawdown or time-normalized head graphs (type curve or straight line matching). Double logarithmic or semilogarithmic graphs are usually used with computer screen displays. Dimensional format refers to the interactive or automatic calibration of calculated and measured time-drawdown or time-normalized head values. Dimensionless calibration consists of finding a best fit of calculated type curve graphs and measured time-drawdown or time-normalized head graphs. During dimensional calibration, differences between calculated and measured time-drawdown or time-normalized head values (residuals) are minimized by varying aquifer conditions, dimensions, and parameters.

There are four aquifer test analysis techniques:

1. Single plot type curve matching with interactive calibration
2. Single plot straight line matching with interactive calibration
3. Composite plot matching with interactive calibration
4. Composite plot automatic parameter estimation

Single plot type curve or straight line matching with interactive calibration precedes composite plot matching with interactive calibration, which is followed by composite plot automatic parameter estimation. *Single plot* refers to the analysis of data for a single well, whereas, *composite plot* refers to the combined analysis of data for all wells. Single plot type curve or straight line matching with interactive calibration techniques guide composite plot matching with interactive calibration techniques and automatic parameter estimation techniques.

Type curve matching is usually best suited for the analysis of observation well data, whereas, straight line matching is best suited for the analysis of pumped or slugged well data. Accentuation of early dimensionless and dimensional time-drawdown data in double-logarithmic graphs facilitates the analysis of well-bore storage, well partial penetration, delayed drainage at the water table under unconfined aquifer conditions, and aquifer boundary impacts.

In the composite plot automatic parameter estimation (nonlinear regression) technique, differences between simulated time-drawdowns or time-normalized heads based on a conceptual model with selected parameter values and measured time-drawdowns or time-normalized heads are minimized using a weighted sum of squared errors objective. An appropriate mathematical modeling equation computer program is run repeatedly while automatically varying the parameter values in a systematic manner from one run to the next until the objective function is minimized. Statistics are provided showing the precision of calculated parameter values.

There is a growing recognition of the importance of including automatic parameter estimation in aquifer test analysis. The expanding use of numerical models and automatic parameter estimation methods in solving inverse problems in hydrogeology is summarized by Carrera et al. (2005). The potential use of automatic parameter estimation regularization and pilot points capabilities of PEST to infer the spatial distribution of parameter heterogeneities is described by Doherty (2003).

Well function or drawdown-head calculation is usually accomplished with software programs such as WTAQ. These programs involve integral and Stehfest algorithm mathematical modeling equations.

There are two types of calibration:

1. Interactive
2. Automatic parameter estimation

Interactive calibration usually precedes and guides automatic parameter estimation. During calibration, the differences between measured and calculated drawdowns (residuals) are minimized. Calibration target windows (acceptable errors) should be predetermined and included in the conceptual model based on information concerning measurement errors and aquifer test conditions. If the interactive calibration is deemed unacceptable, the conceptual model is adjusted, the aquifer test data is reanalyzed, and the calibration process is repeated. Iterative interactive data analysis tasks are:

- Write input-data files
- Run software program
- Read output-data files
- Compare measured and output data
- Revise input-data files
- Rerun software program
- Display data

Input-data files can be written with word processor or spreadsheet software or aquifer test preprocessor software. To write input-data files with a word processor requires the user to become familiar with the content of WTAQ, MODFLOW, and PEST input-data and output files. A preprocessor (graphical user interface [GUI]) automatically creates the input-data files graphically and runs the aquifer test analysis software. The user does not need to see the input-data files or know the commands that run the software until something goes wrong or the user tries to do something out of the ordinary. Then, it is important that the user understand input-data and output-data files and run commands so that the user can track down and resolve the problem. The user must understand which data goes on each file line (record) and in which order (fields). Users can become familiar with WTAQ and MODFLOW input-data files by reading instructions and examples presented by Barlow and Moench (1999) and Andersen (1993). These documents are available at water.usgs.gov/nrp/gwsoftware/ and www.epa.gov/ada/csmos/models/modflow.html.

Regardless of whether preprocessors or word processors are used to create input-data files, conceptual model and computer program solution data must be entered by the user and written to input-data files. Data entry and file management can be quite laborious and tedious. Running software programs can require some knowledge of the disk operating system (DOS) language because software programs such as WTAQ, MODFLOW-96, and PEST are Fortran batch programs that run from a composite model command line containing a redirect file reference in a DOS window.

Model evaluation is an assessment of the reliability, precision, and applicability of calculated aquifer characteristic values. Aquifer test analysis results tend to be approximate and nonunique because test facilities are usually limited, test conditions are usually not ideal, field measurements are usually limited in accuracy and quantity, aquifer test conceptual models and equations seldom completely simulate reality, and observed time-drawdown or time-normalized head values can be duplicated with more than one combination of aquifer parameter values and boundary conditions. Aquifer test analysis results do not always pertain to the entire aquifer thickness and aquifer test domain.

Recent advances in the analysis of aquifer test data are aimed primarily at providing information about spatial variations in hydraulic conductivity and involve large drawdown slug tests, hydraulic tomography, direct-push hydraulic profiling, and spatial weighting coefficients (Frechet kernels). Large drawdown slug tests (see Gonzalo Pulido, HydroQual, Inc. at gpulido@hydroqual.com) are slug tests with large normalized heads greater than 5 m that enhance the estimation of hydraulic parameters with data from observation wells tens of meters from the slugged well. *Hydraulic tomography* (Bohling et al., 2003; Bohling et al., 2002; Butler et al., 1999; and Yeh and Liu, 2000) involves the performance of a series of pumping tests stressing different vertical aquifer intervals with drawdowns measured at multiple observation wells during each test. Data from all tests are analyzed simultaneously to characterize the hydraulic conductivity variation between wells. *Direct-push hydraulic profiling* (Butler et al., 2000) involves performing a series of slug tests in direct-push rods as the rods are driven progressively deeper into the formation. Spatial weighting Frechet kernels provide information about inhomogeneities within the effective volume of influence of time-drawdown data (Knight and Kluitenberg, 2005).

2 Conceptual Model Definition

Unfortunately, conceptual models are rarely properly defined before aquifer test data are analyzed. As a result, analysis often proceeds with an invalid or inadequate understanding of the groundwater flow system. This can lead to confusion and erroneous conclusions about aquifer and confining unit characteristics. Therefore, the first step in aquifer test modeling should be conceptual model definition. In a sense, the conceptual model is the aquifer test modeling database. A conceptual model provides a frame of reference and direction for the entire aquifer test modeling process.

A conceptual model is a space and time representation (approximation) of the groundwater flow system within an aquifer test domain. An aquifer test domain is a cylindrical volume of the groundwater flow system surrounding a pumped or slugged well. The aquifer test domain height is the groundwater flow system thickness, which can be 100 ft or more. A pumping test domain radius can be 500 ft or more under unconfined aquifer conditions and 1000 ft or more under confined nonleaky, leaky, or fissure and block aquifer conditions. The slug test domain height can be 20 ft or more. A slug test domain radius can be 20 ft or more.

A conceptual model organizes, simplifies, and idealizes available information and is based on the:

- Hydrostratigraphic framework of the groundwater system (type, thickness, extent, heterogeneities, and boundaries of hydrostratigraphic units)
- Hydraulic characteristics of hydrostratigraphic units (hydraulic conductivity and storage)
- Pumped or slugged and observation well or piezometer characteristics (such as casing radius, effective radius, pump pipe radius, well-bore storage, and well-bore skin)
- Time dimensions (aquifer test elapsed time in analytical groundwater flow models or the number and lengths of stress periods and time steps in numerical groundwater flow models)
- Groundwater and surface water budgets (pump discharge or slug initial displacement and any recharge)
- Flow system temporal changes (such as water level fluctuations due to atmospheric pressure changes, stream stage changes, or tidal fluctuations)

There are several online pumping test publications illustrating the various methods used in conceptual model definition including the following Web sites:

water.usgs.gov/pubs/pp/pp1629
water.usgs.gov/pubs/wri/wri934008/wri934008_files/wri934008.pdf
pubs.usgs.gov/sir/2005/5233/pdf/sir2005-5233.pdf
ga.water.usgs.gov/pubs/wrir/wrir97–4129/pdf/wrir97-4129.pdf
fl.water.usgs.gov/PDF_files/ofr99_185_broska.pdf
pubs.water.usgs.gov/ofr98294

Conceptual model definition should take into consideration the simplifying assumptions upon which the aquifer test mathematical modeling equations selected for use are based. For example, typical assumptions for analytical Stehfest algorithm aquifer test mathematical modeling equations include:

- Aquifer type (confined nonleaky, confined leaky, confined fissure and block, or unconfined)
- Aquifer is underlain by an impervious unit and can be overlain by a confining unit, which in turn can be overlain by an aquifer with constant or variable head (source unit) or impervious unit
- Aquifer boundaries or discontinuities within the aquifer test domain are straight line demarcations
- Aquifer is homogeneous, isotropic, and of uniform thickness within the aquifer test domain
- No flow occurs in the aquifer test domain prior to pumping
- Pumping rate is constant

The assumptions for numerical finite-difference groundwater flow mathematical modeling equations, such as those incorporated in MODFLOW, are less restrictive and include:

- Continuous groundwater flow system can be replaced by a discretized grid of nodes and associated finite-difference cells (blocks) simulating one or more aquifer and confining unit layers
- Aquifer test time can be discretized into small blocks of variable lengths

Conceptual model definition should also take into consideration the data-input requirements (file or interactive data entry) of the aquifer test software program selected for use. For example, the data-input file for the U.S. Geological Survey analytical pumping test software program, WTAQ, includes the following data-input options and variables:

- Analysis format
- Aquifer type
- Hydraulic characteristics
- Time information

- Well characteristics
- Measured drawdown data
- Program solution variables

Several data-input files are required by the U.S. Geological Survey numerical groundwater flow modeling software program, MODFLOW. These data-input files include the following data-input options and variables:

- Model layers and grid dimensions
- Stress period dimensions
- Boundary type
- Output control
- Layer type
- Hydraulic characteristics
- Well characteristics
- Iteration parameters

Finally, conceptual model definition should take into consideration the calibration (interactive or automatic parameter estimation) data-input requirements of the groundwater flow modeling software program selected for use. These requirements include:

- Interactive calibration target data
- Identification of adjustable parameters for automatic parameter estimation
- Identification of initial, upper, and lower parameter adjustable limits for automatic parameter estimation

DEFINING HYDROSTRATIGRAPHIC UNIT FRAMEWORK

A key element in the definition of a conceptual model is the development of the aquifer test domain hydrostratigraphic unit framework. Hydrostratigraphic units as defined by Maxey (1964) are bodies of rock with considerable lateral extent that act as a reasonably distinct hydrologic system. Thus, hydrostratigraphic units are hydraulically continuous, mappable, and scale-independent entities. Mappable infers that the subsurface can be subdivided according to hydraulic conductivity (Seaber, 1988). A hydrostratigraphic unit may include a formation, part of a formation, or a group of formations. Hydrostratigraphic unit definition is based on chronostratigraphic and lithostratigraphic information and often requires estimation, interpolation, correlation, and extrapolation based on limited information and considerable scientific judgment of a subjective or intuitive nature (see Stone, 1999 and Weight, 2001).

There are many sources of data that can be used to define the aquifer test domain hydrostratigraphic unit framework including:

- Existing literature, cross sections, and maps
- Interviews with previous investigators
- Lithologic descriptions on well logs
- Stratigraphic information
- Depositional history of an area
- Regional head data
- Electric logs
- Aerial photographs
- Drill cuttings submitted by well contractors
- Geophysical logs
- Field reconnaissance
- Core penetration testing
- Outcrops, quarries, and gravel pits

Aquifer test domain hydrostratigraphic unit (layer) interpretations involve the composite study of well logs, cross sections, contour maps, tables, and data summaries. In conceptual model definition, hydrostratigraphic units can be classified as aquifers or confining units. Aquifers can be classified as confined nonleaky, confined leaky, confined fissure and block, or unconfined. Confining units can be classified as impermeable (nonleaky or aquicludes) or leaky (aquitards). Groundwaterflow systems can consist of a single unconfined aquifer or one or more confined aquifers with one or more confining units.

Hydrostratigraphic unit interpretations include the definition of unit (layer) thickness and extent and the delineation of any unit barrier or recharge boundaries or discontinuities and heterogeneities. Several commercial well log, cross section, fence diagram, geochemical graph, and database software programs are available to assist in this process (visit the following Web sites: www.goldensoftware.com, www.scisoftware.com, and www.rockware.com).

DEFINING ANALYTICAL CONCEPTUAL MODELS

Aquifer and confining unit widths, lengths, thicknesses, and hydraulic characteristics must be uniform or their variability predefined, and boundary or discontinuity demarcations must be straight lines in analytical conceptual models to match the assumptions upon which analytical mathematical modeling equations are based. Variable aquifer and confining unit widths, lengths, thicknesses, and hydraulic characteristics must be converted to equivalent dimensions and characteristics. Multiple boundaries or discontinuities must be idealized to fit comparatively elementary geometric forms such as wedges and infinite and semi-infinite rectilinear strips.

In analytical conceptual models, several formations may be lumped into one hydrostratigraphic unit, and one formation may be subdivided into several hydrostratigraphic units, according to the contrast (usually fivefold or more) of hydraulic characteristics. Each hydrostratigraphic unit is a hydraulic entity (layer). Aquifers become two-dimensional horizontal planes (layers) that are sometimes separated by confining units.

The aquifer type should be specified in analytical conceptual models. Common options are confined nonleaky, confined leaky with constant head aquifer above confining unit, confined leaky with impermeable unit above confining unit, confined leaky with water table in confining unit, confined fissure and block with slab-shaped blocks, confined fissure and block with sphere-shaped blocks, unconfined with delayed drainage at water table, and unconfined without delayed drainage at water table as illustrated in Figure 2.1 to Figure 2.7.

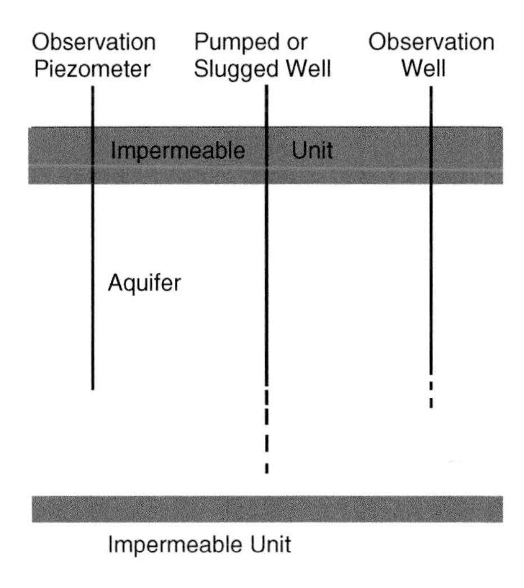

FIGURE 2.1 Confined nonleaky aquifer.

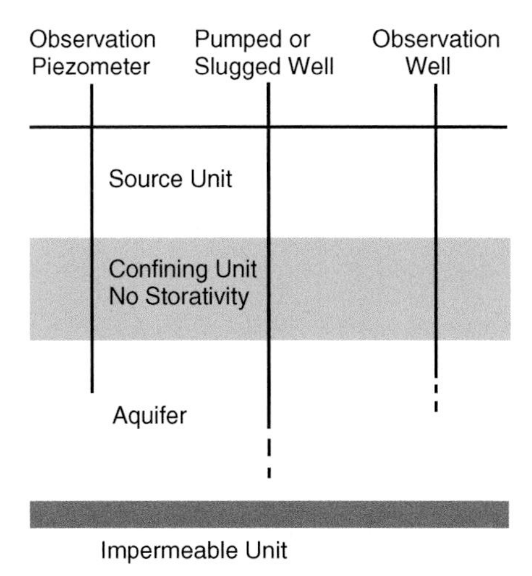

FIGURE 2.2 Confined leaky aquifer with no confining unit storativity and source unit above confining unit.

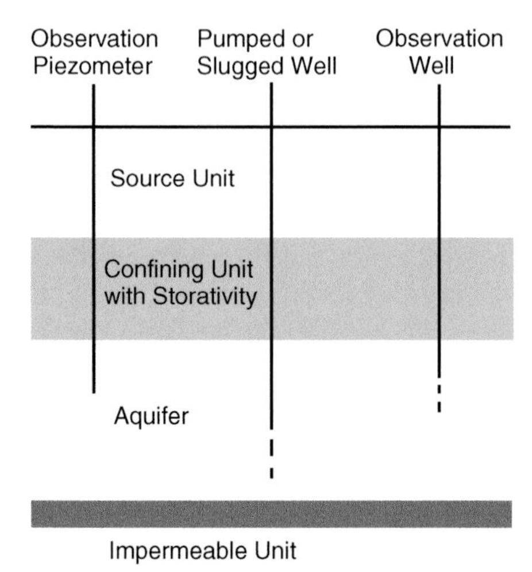

FIGURE 2.3 Confined leaky aquifer with confining unit storativity and source unit above confining unit.

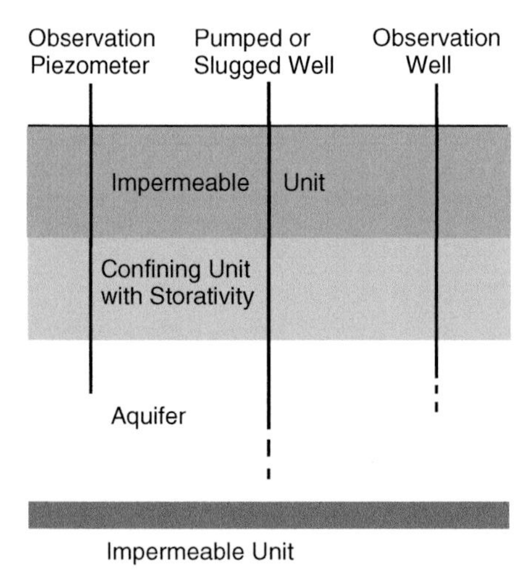

FIGURE 2.4 Confined leaky aquifer with confining unit storativity and impermeable unit above confining unit.

Pumping test time-drawdown derivative graphs (Spane and Wurstner, 1993, p. 816) are useful in verifying initial aquifer type decisions. The time-drawdown derivative curve becomes horizontal under confined nonleaky aquifer conditions as shown in Figure 2.8. Confined leaky conditions cause the time-drawdown derivative curve to plunge downward. Confined fissure and block conditions cause the time-drawdown derivative curve to become horizontal during intermediate time periods and then fall and finally rise during later time periods. The

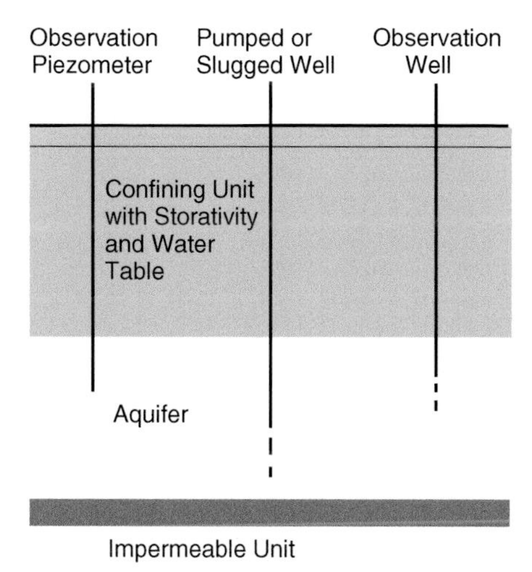

FIGURE 2.5 Confined leaky aquifer with confining unit storativity and water table within confining unit.

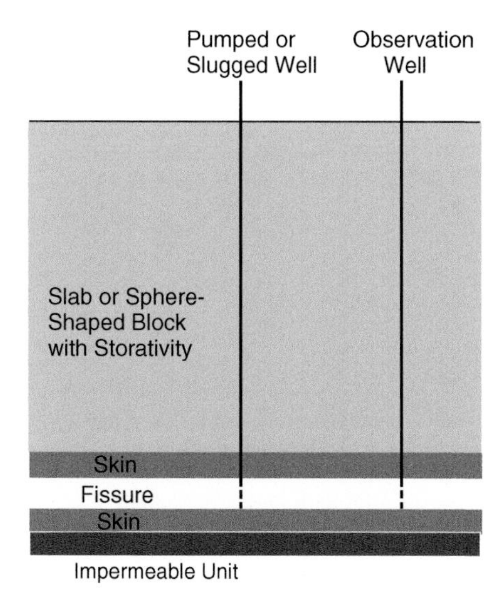

FIGURE 2.6 Confined fissure and block aquifer with slab- or sphere-shaped block with storativity.

time-drawdown derivative curve becomes horizontal during a short intermediate period, falls during another short intermediate period, and rises during a later period under unconfined aquifer conditions. The drawdown derivative curve becomes horizontal with a recharge boundary. A barrier boundary causes the drawdown derivative curve to plunge downward.

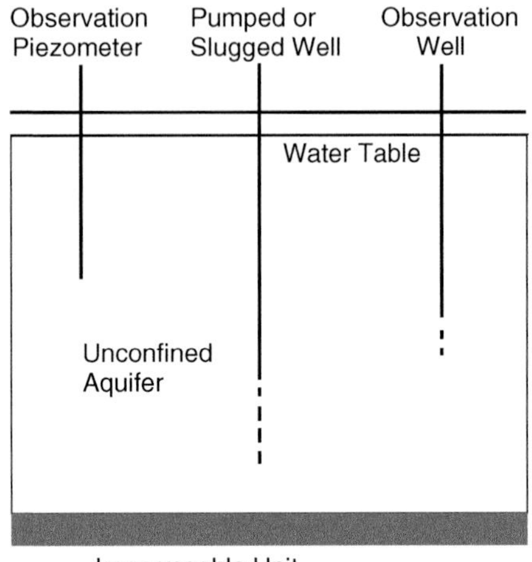

FIGURE 2.7 Unconfined aquifer with or without delayed drainage at water table.

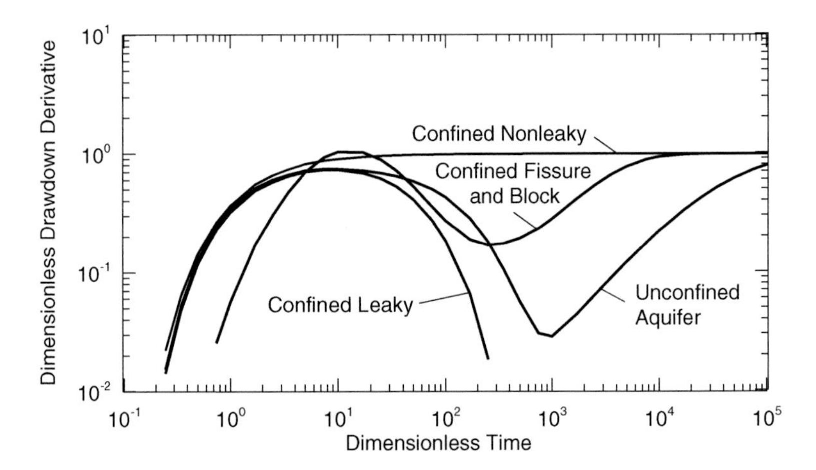

FIGURE 2.8 Aquifer identification with derivative graphs.

HETEROGENEOUS AND VERTICAL ANISOTROPIC CONDITIONS

Heterogeneous aquifer conditions can be defined in an analytical conceptual model with equivalent homogeneous aquifer characteristic values that give the same long-term hydraulic performance as the heterogeneous formations. For example, an aquifer consisting of several (n) horizontal layers, each with different thicknesses and horizontal hydraulic conductivities, can be defined approximately in a conceptual model with an equivalent aquifer single layer.

The equivalent horizontal hydraulic conductivity (K_{eh}) of the aquifer single layer can be calculated with the following equation (Cheng, 2000, p. 42):

$$K_{eh} = 1 / b_{ev} \sum_{i=1}^{n} b_i K_i \tag{2.1}$$

where b_{ev} is the equivalent aquifer single-layer thickness, b_i is the individual aquifer layer thickness, and K_i is an individual aquifer layer horizontal hydraulic conductivity.

The equivalent specific storage (S_{se}) of the aquifer single layer can be calculated with the following equation (Cheng, 2000, p. 44):

$$S_{se} = 1 / b_{ev} \sum_{i=1}^{n} b_i S_{si} \tag{2.2}$$

where b_{ev} is the equivalent aquifer single-layer thickness, b_i is the individual aquifer layer thickness, and S_{si} is an individual aquifer layer-specific storage.

Vertical anisotropy is important near the pumped or slugged well where the effects of well partial penetration are significant. Vertical anisotropy is usually defined in an analytical conceptual model as a ratio of the aquifer vertical hydraulic conductivity divided by the aquifer horizontal hydraulic conductivity. The ratio depends upon the degree of aquifer stratification and is commonly estimated by studying well logs and published tables of ratio values.

A confining unit consisting of several (n) vertical layers, each with different thicknesses and vertical hydraulic conductivities, can be defined approximately in an analytical conceptual model as an equivalent confining unit single layer.

The equivalent vertical hydraulic conductivity (K_{ev}) of the single layer can be calculated with the following equation (Cheng, 2000, p. 43):

$$K'_{ev} = b'_{cev} / \sum_{i=1}^{n} b'_i / K'_i \tag{2.3}$$

where b_{cev} is the equivalent confining unit single-layer thickness, b_i is the individual confining unit layer thickness, and K_i is an individual confining unit layer vertical hydraulic conductivity.

The areal average aquifer transmissivity (T_{av}) and storativity (S_{av}) under heterogeneous conditions can be estimated with the following equations (Raudkivi and Callander, 1976, p. 118):

$$T_{av} = (T_1A_1 + T_2A_2 + \ldots + T_nA_n)/(A_1 + A_2 + \ldots A_n) \tag{2.4}$$

$$S_{av} = (S_1A_1 + S_2A_2 + \ldots + S_nA_n)/(A_1 + A_2 + \ldots A_n) \tag{2.5}$$

where T_n is the aquifer transmissivity in area n, S_n is the aquifer storativity in area n, and A_n is the area underlain by aquifer transmissivity T_n.

HORIZONTAL ANISOTROPIC CONDITIONS

Some sedimentary and fractured aquifers are horizontally anisotropic. The horizontal hydraulic conductivity in one direction may be between 2 and 20 or more times the horizontal hydraulic conductivity in another direction. Drawdown contours around a pumped well in an anisotropic aquifer form concentric ellipses rather than circles, as they would in an isotropic aquifer. Major and minor directions of transmissivity coincide with major and minor ellipse axes. Horizontal anisotropy is defined in numerical conceptual models by varying finite-difference grid cell hydraulic characteristics or by specifying an anisotropy factor other than 1.0.

Horizontal anisotropy is difficult to define in an analytical conceptual model. Horizontal anisotropy aquifer test data can be analyzed with analytical methods derived by Hantush (1966a and 1966b) and Hantush and Thomas (1966) as explained in Kruseman and de Ridder (1994) and Batu (1998). These analytical methods, which will be explained later, involve the postprocessing of horizontally isotropic analytical aquifer test modeling results. Available information concerning the principal directions of horizontal anisotropy should be included in the analytical conceptual model.

WELL PARTIAL PENETRATION

Flow toward fully penetrating wells is horizontal. There are no vertical components of flow. Drawdown in the aquifer with fully penetrating wells varies only with the radial distance from the pumped well and time. Wells with screen lengths less than the aquifer thickness are called *partially penetrating wells*. Flow toward partially penetrating wells is both horizontal and vertical at short distances from the pumped well.

The aquifer vertical hydraulic conductivity is generally less than the aquifer horizontal hydraulic conductivity because of aquifer anisotropy. Consequently, drawdown in the aquifer with partially penetrating wells varies with the length and space position of the screens in the pumped or slugged and observation wells (Hantush, 1964) as well as the radial distance from the pumped well and time. Drawdown in a partially penetrating pumped well can be greater or less than the drawdown in a fully penetrating pumped well.

The effects of well partial penetration increase during early portions of a pumping test and then stabilize. Partial penetration effects may be negative or positive depending on well geometry. For example, if the pumped and observation wells are both open in either the top or bottom portion of the aquifer, the drawdown in an observation well is greater than it would be with fully penetrating wells. If the pumped well is open to the top of the aquifer and an observation well is open to the bottom of the aquifer, or vice versa, the drawdown in the observation well is less than for fully penetrating conditions (see Reed, 1980). Drawdowns measured in a piezometer are those that occur in the aquifer at the center of piezometer.

The distance from the pumped or slugged well beyond which the effects of well partial penetration are negligible (r_b) is estimated with the following equation (Hantush, 1964, p. 351):

$$r_b = 1.5b(K_h/K_v)^{0.5} \tag{2.6}$$

where b is the aquifer thickness, K_h is the aquifer horizontal hydraulic conductivity, and K_v is the aquifer vertical hydraulic conductivity.

Partial penetration effect distances increase with aquifer thickness and horizontal-to-vertical hydraulic conductivity ratio from less than 10 to several hundred feet from a pumped or slugged well.

Analytical conceptual models should define well penetration dimensions so that the effects of partially penetrating wells can be simulated.

WELLBORE STORAGE

During very early portions of a pumping test, part of the water discharged from a well is derived from water stored in the well casing (*wellbore storage*) and is not withdrawn from the aquifer system (*aquifer flow rate*). The aquifer flow rate and the associated time rate of drawdown are less at very early times than they are at later times because of wellbore storage effects. If the pumped well has a finite diameter and wellbore storage is appreciable, the discharge rate is the sum of the aquifer flow rate and the rate of wellbore storage depletion.

The aquifer flow rate increases exponentially with time toward the discharge rate and the wellbore storage depletion rate decreases in a like manner to zero (Streltsova, 1988, pp. 49–55). Also, during very early portions of a slug test, the slug displacement in the aquifer is less than the slug displacement within the slugged well because of wellbore storage.

Wellbore storage effects usually range in duration from the first few minutes of an aquifer test with moderate to high (> 1000 ft^2/day) transmissivities and small (< 0.5 ft) well radii to several hours or days with lower transmissivities or larger well radii. Wellbore storage effects decrease with time (Fenske, 1977). Wellbore storage effects increase as the distance from the pumped or slugged well and storativity decrease, and the pumped or slugged well radius increases.

The duration (t_s) of wellbore storage impacts can be estimated with one of the following equations (Papadopulos and Cooper, 1967, p. 242):

$$t_s = 250(r_c^2 - r_p^2)/T \text{ (for pumped well)} \tag{2.7}$$

$$t_s = 2500(r_c^2 - r_p^2)/T \text{ (for observation well)} \tag{2.8}$$

where T is aquifer transmissivity, r_c is the pumped well casing radius, and r_p is the pump pipe radius. The (r_c^2 r_p^2) term allows for the presence of a column pipe or other tubing that reduces the cross-sectional area of the pumped well in the vicinity of changing water levels. If the radii in the term are not known, their values can be estimated with data in Table 2.1 and Table 2.2 (based on data in Driscoll, 1986, pp. 415 and 1028).

Pumped wellbore storage effects are simulated in WTAQ with a dimensionless pumped wellbore storage parameter W_d defined as follows (Barlow and Moench, 1999, p. 11):

$$W_d = (\pi r_{ce}^2)/[2\pi r_w^2 S_s(z_{pl} - z_{pd})] \quad (2.9)$$

where r_w is the effective radius of the pumped well, r_{ce} is the effective radius of the pumped well casing in the interval where water levels are changing (allows for the presence of a column pipe or other tubing that might reduce the cross-sectional area of the pumped well casing) outside the radius of the pumped or slugged well screen, S_s is the aquifer specific storage, z_{pl} is the depth below the

TABLE 2.1
Typical Pumped Well Diameters

Discharge Rate (ft³/day)	Diameter (ft)
< 19251	0.50
14439–33690	0.67
28877–67380	0.83
57754–134759	1.00
96257–192513	1.17
154011–346524	1.33
231016–577540	1.67

TABLE 2.2
Typical Pump Pipe Diameters

Discharge Rate (ft³/day)	Diameter (ft)
< 4813	0.10
9626	0.17
19,251	0.25
38,503	0.33
77,005	0.50
115,508	0.67
192,513	0.83
385,027	1.17
577,540	1.33

top of aquifer or initial water table to the bottom of the screened interval of the pumped well, and z_{pd} is the depth below the top of the aquifer or initial water table to the top of the screened interval of the pumped well.

Observation wells contain a significant quantity of stored water in their casings or open holes. During early portions of aquifer tests, rapid drawdown in the aquifer may not be accurately reflected by measured drawdowns in the observation well because of the finite time it takes to dissipate stored water and come into equilibrium with drawdown in the aquifer (Moench et al., 2001, p. 12). Delayed response of an observation well is simulated in WTAQ with a dimensionless observation well delayed response parameter W_{dp} defined as follows (Barlow and Moench,1999, p.12):

$$W_{dp} = (\pi r_{co}^2)/(2\pi r_w^2 S_s F') \quad (2.10)$$

where r_{co} is the inside radius of the observation well in the interval where water levels are changing, r_w is the effective radius of the pumped well, S_s is the aquifer-specific storage, and F' is a shape factor defined by Hvorslev (1951, case 8).

$$F' = L_s/\ln[x + (1 + x^2)^{0.5}] \quad (2.11)$$

where

$$x = (mL_s)/(2r_{co}) \quad (2.12)$$

$$m = (K_h/K_v)^{0.5} \quad (2.13)$$

r_{co} is the inside radius of the observation well in the interval where water levels are changing, L_s is the length of the screened interval of the observation well, K_h is the aquifer horizontal hydraulic conductivity, and K_v is the aquifer vertical hydraulic conductivity.

Delayed response in an observation piezometer cannot be simulated with the above equations because the piezometer has a zero screen length. However, a piezometer can be simulated as a finite-screen length observation well with a very small radius.

Analytical conceptual models should define well dimensions so that wellbore storage in the pumped or slugged well and delayed response in observation wells can be simulated.

WELLBORE SKIN

A thin wellbore skin can be present at the interface between the pumped well screen and the aquifer (see Moench, 1985). Wellbore skin may be less than or greater than that of the aquifer due to well construction and development practices. The wellbore skin is assumed to have hydraulic conductivity but negligible storage capacity. The hydraulic conductivity is assumed to be constant during an aquifer test. Pumped wellbore skin is simulated in WTAQ with the dimensionless pumped wellbore skin parameter *SW* and the following equation (Barlow and Moench, 1999, p. 11):

$$SW = (K_h d_s)/(K_s r_w) \tag{2.14}$$

where d_s is the thickness of the wellbore skin, K_h is the aquifer horizontal hydraulic conductivity, K_s is the hydraulic conductivity of the wellbore skin, and r_w is the pumped wellbore effective radius.

In the absence of wellbore skin, *SW* is set to 0.0. When pumped wellbore skin with a hydraulic conductivity less than that of the aquifer is simulated but wellbore storage is not, drawdowns in the aquifer will not be affected by the wellbore skin. However, drawdowns in the pumped well will be greater than they would be in the absence of the wellbore skin. When both wellbore storage and skin are simulated, early time drawdowns at observation wells and piezometers will be delayed relative to early time drawdowns that occur in the absence of wellbore skin. Drawdowns at the pumped well will be greater than they would be in the absence of wellbore skin (Barlow and Moench, 1999, p. 11).

Effects of any wellbore skin should be considered in estimating the pumped well effective radius (Moench et al., 2001, p. 18). If the wellbore skin is less permeable than the aquifer, drawdown in the pumped well is increased (the effective radius decreases) and there is an apparent increase in wellbore storage, which reduces drawdowns in the aquifer at early times. If the well skin is more permeable than the aquifer, drawdown in the pumped well is decreased (the effective radius increases) and there is an apparent decrease in wellbore storage, which increases drawdowns in the aquifer at early times.

The pumped well effective radius can be estimated by calculating pumped well drawdowns for a selected time based on aquifer parameter values estimated with observation well data and several trial effective radius values. Calculated drawdowns are compared with the measured drawdown for the selected time. The trial effective radius that results in a match between calculated and measured drawdowns is assigned to the pumped well.

Wellbore skin dimensions and hydraulic characteristic values should be defined in analytical conceptual models so that the effects of wellbore skin can be simulated.

UNCONFINED AQUIFERS

In unconfined aquifers, pumping test time-drawdown data usually show a typical S-shape composed of three distinct segments: a steep early time segment, a flat intermediate time segment, and a relatively steep late time segment (see Neuman, 1972). The first segment covers a brief period often only a few minutes in length during which the unconfined aquifer reacts in the same way as a confined aquifer.

The water discharged from the well is derived from aquifer storage by the expansion of the water and the compaction of the aquifer.

The second segment, which can range in length from several minutes to days, mainly reflects the impact of gradual drainage of saturated interstices and the delayed drainage of the unsaturated zone within the cone of depression created during the first segment. If the effects of delayed drainage of the unsaturated zone are not taken into account, the specific yield estimated with aquifer test data can be significantly underestimated (Moench et al., 2001, p.1)

The third segment reflects a period during which the water discharged from the well is derived both from gradual gravity drainage of interstices and the expansion of the water and the compaction of the aquifer as the cone of depression expands continuously.

The duration (t_d) of delayed gravity drainage under unconfined aquifer conditions can be estimated with the following equation (see Boulton, 1954a):

$$t_d = 2bS_{wt}/K_v \quad (2.15)$$

where b is the aquifer thickness, S_{wt} is the aquifer-specific yield, and K_v is the aquifer vertical hydraulic conductivity.

Delayed gravity drainage impacts increase with aquifer thickness and decrease as the vertical hydraulic conductivity increases from less than one hour to several days. Drainage from the unsaturated zone above the water table can be treated as either instantaneous or delayed in WTAQ (Barlow and Moench, 1999). Instantaneous drainage is simulated with a single drainage constant (1.0×10^9). Delayed drainage is simulated with a finite (3–5) series of exponential drainage constants such as 0.0346 min^1, 25.92 min^1, and 31680 min^1 (see Moench et al., 2001, p. 45).

Information concerning delayed drainage at the water table and any delayed drainage constant initial, upper limit, and lower limit values should be defined in the analytical conceptual model.

HYDROGEOLOGIC BOUNDARIES

The existence of hydrogeologic boundaries (full or partial barrier or recharge) can limit the continuity of an aquifer in one or more directions. Partial barrier or recharge boundaries are called *discontinuities*. Pumping test data will show the impacts of a full boundary when aquifer transmissivity in the immediate vicinity of the pumping well is ten times greater than or one tenth less than the aquifer transmissivity at some distance from the pumping well (Fenske, 1984, pp. 131–132).

A full barrier boundary is defined as a line (streamline) across which there is no flow, and it may consist of folds, faults, or relatively impervious deposits such as shale or clay. A full recharge boundary is defined as a line (equipotential) along which there is no drawdown, and it may consist of increased aquifer transmissivity or streams, lakes, and other surface water bodies hydraulically

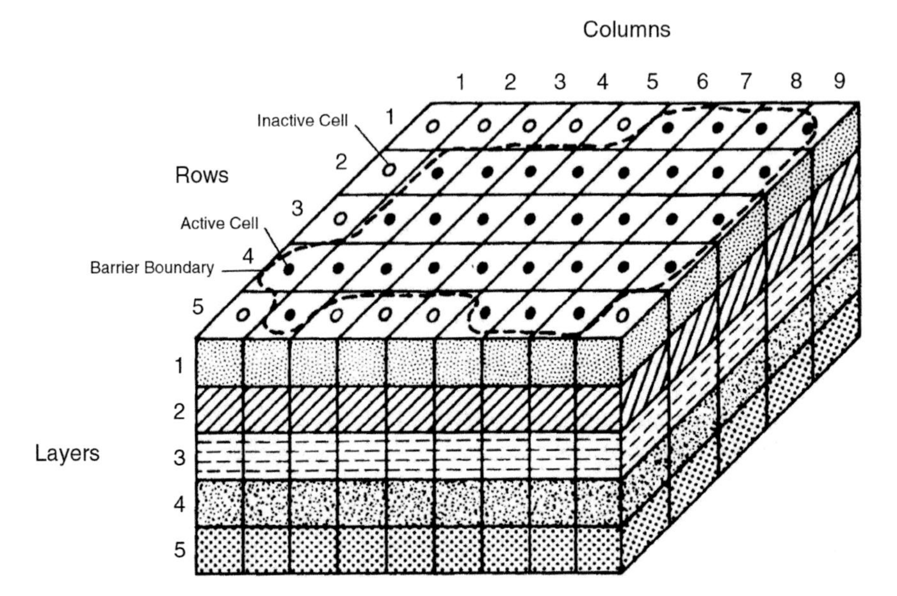

FIGURE 2.9 Discretized numerical model (after McDonald and Harbaugh, 1988, *U.S. Geological Survey Techniques of Water-Resources Investigations*, Book 6, Chap. A-1).

connected to the aquifer. Most full hydrogeologic boundaries are not clear-cut straight-line features but are irregular in shape and extent. However, complicated full boundaries must be simulated with straight-line demarcations in analytical mathematical modeling equations.

Finite groundwater system conditions within the aquifer test domain can be simulated with the image well theory to be explained later by replacing hydrogeologic boundaries with imaginary wells that produce the same disturbing effects as the boundaries (Ferris et al., 1962). Analytical conceptual models should define the locations of any image wells associated with finite aquifer conditions.

DEFINING NUMERICAL CONCEPTUAL MODELS

Conceptual models for use with groundwater flow numerical models, such as the U.S. Geological Survey program MODFLOW (McDonald and Harbaugh, 1988), consist of a discretized grid of layers, columns, and rows defined in a basic data-input file as illustrated in Figure 2.9. Kolm (1993) presents an integrated, stepwise approach for conceptualizing and characterizing hydrologic systems thereby leading to the description of a numerical model simulating the study area. Each layer's type is identified by the term LAYCON in the Block-Centered Flow (BCF) data-input file. Possible layer types are:

- Confined—transmissivity and storativity of the layer are constant for the entire simulation.

- Unconfined—transmissivity of the layer varies. It is calculated from the saturated thickness and hydraulic conductivity. The storativity is constant. Valid only for Layer 1.
- Confined/Unconfined—transmissivity of the layer is constant. The storativity may alternate between confined and unconfined values. Vertical leakage from above is limited if the layer desaturates.
- Confined/Unconfined—transmissivity of the layer varies. It is calculated from the saturated thickness and hydraulic conductivity. The storativity may alternate between confined and unconfined values. Vertical leakage from above is limited if the aquifer desaturates.

The horizontal grid for areal pumping test simulations typically consists of 50 rows and 50 columns within the pumping test domain. The grid is aligned appropriately with respect to the pumping test domain and any aquifer boundaries or heterogeneities. The pumped well is usually located in the center block of the grid. Typically, square blocks of the grid have spacing of 100 ft or less within a radius of 1000 ft of the pumped well. The grid spacing is expanded beyond 1000 ft out to the grid borders (boundaries) by increasing nodal spacing no more than 1.5 times the previous nodal spacing. The grid can be designed so there is negligible drawdown at the grid borders that do not represent boundaries. The grid can be designed so that grid nodes are at the location of observation wells to avoid spatial interpolation of output files during calibration.

The grid for cylindrical well simulations is turned on its side and typically consists of several rows and 40 columns radially spaced with a multiplier of 1.5. The grid starts at the top row (layer thickness) and ends at the bottom row. Layer thickness for the top and bottom rows is one half the adjoining row thickness. The pumped well is located in Row 1 and there is negligible drawdown along Column 40, which is a constant-head boundary. Under unconfined conditions, there are usually 11 or more rows in order to simulate slow gravity drainage under unconfined aquifer conditions.

The effects of partially penetrating wells are simulated in numerical conceptual models by specifying the layers in which wells are open. The finite-difference grid and the layers in which wells are located should be defined in numerical conceptual models.

CONFINING UNITS

Analytical aquifer test modeling software is based on an explicit representation of a confining unit, whereas, confining units are not explicitly represented in MODFLOW. Instead, a standard part of MODFLOW (BCF package) indirectly simulates steady-state confining unit leakage without confining unit storage changes by means of a vertical leakance (vertical confining unit hydraulic conductivity divided by confining unit thickness) term known as VCONT at each finite-difference grid node. The source of water to the confined leaky aquifer may be another confined aquifer or an unconfined aquifer. However, it is assumed that the head in the source unit is constant, there is no release of water from storage within the confining unit, and flow in confining units is vertical (horizontal flow in confining units is negligible). Thus, confining unit storativity is not simulated in MODFLOW because most models do not need to simulate this condition. However, aquifer test conceptual models must take into consideration confining unit storativity so that very early time-drawdown data can be accurately calculated (see Hantush, 1964).

The time period during which confining unit storativity effects are appreciable (t_{cs}) can be estimated with the following equation (Trescott et al., 1976):

$$t_{cs} = S'b'/(2K') \tag{2.16}$$

where S' is the confining unit storativity, b' is the confining unit thickness, and K' is the confining unit vertical hydraulic conductivity.

Transient confining unit leakage can be simulated in aquifer test conceptual models by using several layers (usually less than 20) with assigned values of specific storage and leakance to represent the confining unit (McDonald and Harbaugh, 1988). Hydrologically equivalent confining unit-specific storage and leakance values are assigned to individual layers in proportion to conceptual model dimensions in the simulation of transient confining unit leakage with layers. Very small transmissivities are assigned to the layers so that the flow in the confining unit is essentially vertical.

The number of layers is determined by the method of successive approximations wherein the results of simulations with a given number of layers are compared with analytical solutions. Use of one layer to simulate transient confining unit leakage results in the vertical distribution of head in the confining unit being approximated with one head value at the center of the confining unit. The addition of layers to represent the confining unit adds detail to the approximation of the head value at the center of the confining unit. A source unit with a constant head (such as a surface water body) is simulated by assigning a large (usually 0.2 to 0.5) specific yield to each source unit finite-difference grid node.

The TLK1 (Leake et al., 1994) MODFLOW package can be used to take confining unit transient leakage into consideration without the use of additional model layers to simulate a confining unit. The confining unit must be bounded above and below by model layers in which the head is calculated or specified. For a confining unit that pinches out, transient equations are used where the confining unit exists and VCONT terms are used where the confining unit is absent. Specific storage is assumed to be constant. The VCONT terms for layers surrounding a confining unit are set to 0.0 in the BCF package.

When a transient leakage parameter at a node is set to zero or less, TLK1 does not carry out transient leakage calculations at that node. Instead, leakage is calculated using the VCONT value for that node in the BCF package. TLK1 cannot simulate transient leakage in a confining unit that is bounded on the top or bottom by an impermeable boundary nor a situation where the water table is within the confining unit. The wetting capability should not be used for any model layers that connect to a confining unit that is being simulated with the TLK1 package. The numerical conceptual model should contain information about confining unit simulation.

UNCONFINED AQUIFERS

Gradual drainage of saturated interstices is not simulated in MODFLOW because most models do not need to simulate this short time condition. However, the pumping test conceptual model definition should take into consideration gradual drainage of saturated interstices so that very early time-drawdown data can be accurately calculated. The simulation can be accomplished by using fine discretization in time and space in an areal or cylindrical well simulation (Reilly and Harbaugh, 1993a, pp. 489–494).

The unconfined aquifer is subdivided into several (usually 10 or more) layers, especially in the upper parts of the aquifer, within which gradual drainage of interstices occurs depending upon the desired simulation accuracy. Confined primary storage coefficients are assigned to all layers except the uppermost layers. Aquifer-specific yields are assigned to the secondary storage coefficients for the uppermost layers. The optimal discretization of layers can be determined by varying the number of layers, simulating time drawdown in aquifer systems with uniform parameter values, and comparing MODFLOW results and time-drawdown values calculated with analytical Stehfest algorithm modeling equations. Delayed drainage from the unsaturated zone above the water table described by Moench et al. (2001) is not simulated in MODFLOW. The numerical conceptual model should contain information about the simulation of unconfined aquifers.

WELLBORE STORAGE

A pumped well is simulated in MODFLOW by imposing a discharge rate on a grid block node. The grid block node is usually much larger than the pumped well diameter. The drawdown calculated by MODFLOW at the pumped well node is an average drawdown for the grid block not the drawdown in the pumped well (Beljin, 1987, pp. 340–351). The concept of equivalent well block radius with a square (uniform) or rectangular (variable) grid and anisotropic aquifer hydraulic conductivity (Peaceman, 1983, pp. 531–543) is used to estimate the drawdown in the pumped well based on the drawdown calculated by MODFLOW. Wellbore skin is not simulated in conceptual models for numerical groundwater flow software such as MODFLOW.

Pumped wellbore storage is not simulated in MODFLOW because most models do not need to simulate this short-term well characteristic. However, pumping test conceptual models should consider pumped wellbore storage so that very early time-drawdown data can be accurately generated. Pumped wellbore storage with areal flow to a well can be simulated by estimating aquifer flow rates and setting MODFLOW discharge rates equal to aquifer flow rates.

Aquifer flow rates can be estimated by analytically calculating drawdowns at the pumped well for selected elapsed times with and without pumped wellbore storage using average aquifer parameter values for the pumping test site and a constant discharge rate. The constant discharge rate is multiplied by ratios of drawdowns with and without pumped wellbore storage to estimate aquifer flow rates. Drawdowns with and without pumped wellbore storage can be calculated with the U.S. Geological Survey software program WTAQ.

Pumped well wellbore storage with cylindrical flow to a well can be simulated by assigning a very large (effectively infinite) radial conductance between nodes that represent the well at the well screen, zero radial conductance between nodes representing the well where the well is cased, a large (effectively infinite) vertical conductance inside the pumped well, and a storage capacity for the topmost node in the well (representing the free surface) that corresponds to a unit value of specific yield.

Observation well delayed response also is not simulated in MODFLOW because most models do not need to simulate this well characteristic. However, some pumping test models must simulate observation well delayed response so that early time-drawdown data can be calculated. The effects of observation well delayed response can be simulated by analytically calculating drawdowns at an observation point for selected elapsed times with and without observation well delayed response using average aquifer parameter values for the pumping test site and a constant discharge rate. Differences in drawdowns with and without observation well delayed response are subtracted from drawdowns previously calculated with MODFLOW and variable discharge rates. Drawdowns with and without observation well delayed response can be calculated with the U.S. Geological Survey software program WTAQ. The numerical conceptual model should contain information about wellbore storage simulation.

HYDROGEOLOGIC BOUNDARIES

Finite groundwater system conditions within the aquifer test domain are simulated in MODFLOW with constant-head and no flow cells as illustrated in Figure 2.9. The following three types of cells can be designated in the IBOUND arrays within the MODFLOW basic data-input file:

- Constant head
- Inactive (no flow)
- Variable head (active)

Actual groundwater boundaries are always of irregular shape, whereas, boundary definition in numerical conceptual models is always rectangular in shape. The numerical conceptual model should contain an IBOUND array.

QUANTIFYING HYDRAULIC CHARACTERISTICS

Initial, upper limit, and lower limit quantitative estimates of aquifer, confining unit, and streambed characteristics should be defined in the conceptual model. Ranges of conceptual model hydraulic characteristics (horizontal and vertical

TABLE 2.3
Typical Horizontal Hydraulic Conductivity Ranges

Deposit	Conductivity (ft/day)
Gravel	100–4000
Basalt	0.0000001–6000
Limestone	0.0003–3000
Sand and gravel	30–700
Sand	3–400
Sand, quick	7–100
Sand, dune	14–40
Peat	1–40
Sandstone	0.0001–7
Loess	0.000002–7
Clay	0.000002–0.3
Silt	0.0001–3
Till	0.000001–3
Shale	0.000002–0.02
Quartzite	0.0005–1
Greenstone	0.01–2
Rhyolite	0.1–0.3
Schist	0.001–0.3
Dolomite	0.00001–300
Gneiss	1E-5–30
Granite	1E-8–300
Tuffs	0.000001–30
Chalk	0.001–3
Coal	0.1–100
Salt	1E-9–1E-7

hydraulic conductivity, vertical stratification, storativity, and specific yield) can be roughly quantified based on hydrogeologic data, laboratory test data, logic, judgment, and the information in Table 2.3 to Table 2.7 (based on data in Polubarinova-Kochina, 1962; Rasmussen, 1964; Davis and DeWiest, 1966; Davis, 1969; Morris and Johnson, 1967; Walton, 1970; Domenico, 1972; Lohman, 1972; Freeze and Cherry, 1979; de Marsily, 1986; Walton, 1991, pp. 414–416; Spitz and Moreno, 1996, pp. 341–354; and Halford and Kuniansky, 2002).

In addition, measured specific capacity Q/s_m data for fully penetrating wells or the specific capacity data for partially penetrating wells (Q/s_{pp}) can be used to estimate the order of magnitude of aquifer transmissivity. If the well partially penetrates the aquifer, the specific capacity with partial penetration Q/s_{pp} can be adjusted to reflect fully penetrating well conditions (Q/s_m) with the following equation (Kozeny, 1933, pp. 88–116):

$$(Q/s_m) = (Q/s_{pp})\ L\ [1 + 7(MN)^{0.5}] \tag{2.17}$$

TABLE 2.4
Typical Vertical Hydraulic Conductivity Ranges

Location	Deposit	Conductivity (ft/day)
Joliet, IL	Sand, gravel, and clay	0.1363
Mattoon, IL	Clay, sand, and gravel	0.0842
Beecher, IL	Clay, sand, and gravel	0.0334
Assumption, IL	Clay, sand, and gravel	0.0254
Dieterich, IL	Clay, sand, and gravel	0.0134
Winchester, IL	Clay, sand, and gravel	0.0107
Oxnard, CA	Silty and sandy clay	0.0080
Arcola, IL	Clay, sand, and gravel	0.0053
Woodstock, IL	Clay, sand, and gravel	0.0013
Libertyville, IL	Dolomite	0.001203
West Chicago, IL	Dolomite	0.000668
Delaware Canal	Clay with sand lenses	0.000174
Chicago, IL	Dolomitic shale	0.0000066

TABLE 2.5
Sample Anisotropy Ranges

Deposit	P_h (ft/day)	P_h/P_v
Sand with some gravel	340	2/1
Sand	34	3/1
Sand and gravel	470	3/1
Sand with some silt	340	3/1
Sand with some gravel	450	3/1
Sand and gravel	570	10/1
Sand with some gravel	48	10/1
Sand and gravel	406	11/1
Sand	220	15/1
Sand and gravel	206	20/1
Sand, limestone, sandstone	481	26/1
Sand with gravel and clay	240	50/1

where $L = L_s/b$, $M = r_w/(2bL)$, $N = \cos(3.1416L/2)$, Q is the discharge rate, s_m is the measured drawdown, L_s is the length of screen or open hole, b is the aquifer thickness, and r_w is the pumped well radius.

Aquifer transmissivity is estimated with the following equation (see Driscoll, 1986):

$$T = F_{sc}(Q/s_m) \tag{2.18}$$

where Q is the pumped well discharge rate, s_m is the measured drawdown, F_{sc} = 1.4 for confined nonleaky, F_{sc} = 1.0 for confined leaky, and F_{sc} = 0.8 for unconfined aquifer conditions, m is aquifer thickness.

Horizontal hydraulic conductivities of aquifers can also be estimated from quantitative interpretation of geophysical logs using equations relating porosity to hydraulic conductivity developed by Jorgensen (1988) and modified by Paillet et al. (1990).

TABLE 2.6
Typical Storativity Ranges

Deposit	Storativity, Dimensionless
Clay	0.00028–0.0078
Sand	0.000039–0.001
Gravel	0.00001–0.0001
Rock	0.000001–0.0001

TABLE 2.7
Typical Specific Yield Ranges

Deposit	Specific Yield, Dimensionless
Peat	0.30–0.50
Sand, dune	0.30–0.40
Sand, coarse	0.20–0.35
Sand, gravelly	0.20–0.35
Gravel, fine	0.20–0.35
Gravel, coarse	0.12–0.25
Gravel, medium	0.15–0.25
Loess	0.15–0.35
Sand, medium	0.15–0.32
Sand, fine	0.10–0.28
Igneous, weathered	0.20–0.30
Sandstone	0.10–0.40
Sand and gravel	0.15–0.30
Silt	0.03–0.19
Clay, sandy	0.03–0.12
Clay	0.01–0.05
Volcanic, tuff	0.02–0.35
Shale	0.01–0.40
Siltstone	0.01–0.35
Limestone	0.01–0.25
Sandstone	0.02–0.41
Schist, weathered	0.06–0.21
Tuff	0.02–0.47
Till	0.05–0.20

Geostatistical correlation techniques that can be applied to the quatification of hydraulic characteristics are presented by Deutsch (2002). These techniques involve correlations based on outcrops and densely drilled similar areas, knowledge of geological processes, vertical and horizontal trends, and relationships between parameters such as porosity and hydraulic conductivity.

DEFINING WELL CHARACTERISTICS

Data-input files for WTAQ contain the following well characteristic information, which should be defined in the conceptual model:

- Effective radius of pumped or slugged well
- Casing radius of pumped or slugged well
- Pumped pipe radius of pumped well
- Type of pumped or slugged well — fully penetrating or partially penetrating
- Depth below top of aquifer or initial water table to top of screened interval of pumped or slugged well
- Depth below top of aquifer or initial water table to bottom of screened interval of pumped or slugged well
- Option for wellbore storage (no wellbore storage or wellbore storage)
- Pumped or slugged wellbore skin thickness
- Pumped or slugged wellbore skin hydraulic conductivity
- Measured time drawdown or head in pumped or slugged well
- Number of observation wells
- Radial distance from pumped or slugged well to observation well or piezometer
- Casing radius of observation well
- Type of observation well — fully penetrating, partially penetrating, or piezometer
- Depth below top of aquifer or initial water table to top of screened interval of observation well
- Depth below top of aquifer or initial water table to top of screened interval of observation well
- Depth below top of aquifer or initial water table to center of piezometer
- Option for delayed observation well response — no delayed response or delayed response
- Measured time drawdown or head in each observation well or piezometer

Well characteristic information required for numerical aquifer test modeling software, such as MODFLOW, which should be defined in the conceptual model include:

- Effective radius of pumped well
- Casing radius of pumped well
- Pumped pipe radius of pumped well

- Measured time drawdown or head in pumped well
- Number of observation wells
- Radial distance from pumped well to observation well
- Casing radius of observation well
- Measured time drawdown or head in each observation well

DEFINING AQUIFER TEST DOMAIN RADIUS

The estimated aquifer test domain radius should be defined in the conceptual model. The aquifer test domain radius can be defined as the radius around the well such that 99% of water withdrawn from the well is produced within the radius. The pumping test domain radius (r_r) in a homogeneous aquifer can be estimated with the following equation (set $u = 5$ in the Theis equation):

$$r_r = (20Tt/S_a)^{0.5} \tag{2.19}$$

where T is the aquifer transmissivity, t is the elapsed time, S_a is the storativity under confined nonleaky aquifer conditions, S_a is the specific yield under unconfined aquifer conditions, or $S_a = 0.005$ under confined leaky aquifer conditions.

The slug test domain radius in a homogeneous aquifer depends on the accuracy of the pressure or head recording devices, well dimensions, and aquifer parameters. Assuming a down hole pressure transducer sensor resolution of 1% of the initial head disturbance, Guyonnet et al. (1993, p. 629 and 633) derived the following equation for calculating the approximate aquifer test domain radius of a slug test:

$$r_{rs} = 8.37r_w[r_c^2/(2r_w^2 S)]^{0.495} \tag{2.20}$$

where r_{rs} is the slug test domain radius, r_w is the effective radius of the slugged well, r_c is the casing radius of the slugged well, and S is the storativity of the aquifer.

DEFINING TIME DIMENSIONS

Data-input files for WTAQ require that the conceptual model contain the following time information:

- Time specification option — log-cycle or user-specified times
- Log-cycle times — largest value of time, number of logarithmic cycles on time scale, and number of equally spaced times per logarithmic cycle for which drawdown will be calculated
- User-specified times — number of user-specified times at which drawdown will be calculated for the pumped well, time and measured drawdown data for pumped well, number of user-specified times at which drawdown will be calculated for each observation well, and time and measured drawdown data for each observation well

Time is discretized in MODFLOW. Simulation time is divided into stress periods defined as time intervals during which stresses are constant. In turn, stress periods are divided into time steps. Within each stress period, the time steps form a geometric progression. MODFLOW data-input files require that the conceptual model contain the following information:

- Lengths of each stress period
- Number of time steps into which each stress period is divided
- Time step multiplier (ratio of the length of each time step to that of the preceding time step)

Several stress periods each with a single time step and a 1.0 time step multiplier are typically specified for pumping test analysis. For simulation of confined aquifer conditions, time is typically subdivided into 36 stress periods, each with a specified aquifer flow rate and logarithmic spaced time lengths ranging from 0.0002 to 0.10 day. For simulation of unconfined aquifer conditions, time is typically more finely subdivided than for confined aquifer conditions into 45 stress periods, each with a different aquifer flow rate and logarithmic spaced lengths ranging from .00002 to 0.1 day. Optimum logarithmic stress period lengths can best be determined with the aid of a semilogarithmic graph.

GROUNDWATER AND SURFACE WATER BUDGETS

During an aquifer test, the groundwater budget usually consists of the constant discharge from a pumped well or the initial displacement in a slugged well. The data-input file for WTAQ contains a single value for the pumping rate of the well. The data-input file MODFLOW contains variable pumping rates to simulate wellbore storage.

The surface water budget can also contain recharge from a surface body of water such as a stream. When a well near a stream hydraulically connected to an aquifer is pumped, groundwater levels are lowered below the surface of the stream and the aquifer is recharged by the influent seepage of surface water. The cone of depression grows until induced streambed infiltration balances discharge. The cone of depression may expand only partway or across and beyond the streambed depending on the streambed dimensions and hydraulic conductivity.

The effects of induced infiltration can be simulated in analytical conceptual models by replacing the stream with a recharging image well for the purpose of estimating the aquifer hydraulic conductivity, storativity, and specific yield. Prior to the pumping test, groundwater level changes in an observation well near the stream due to stream stage changes can be measured. This data can be used to determine the streambed leakance.

The following information is required to simulate induced infiltration from a stream using analytical mathematical modeling equations and should be defined in the conceptual model:

- Groundwater level change in an observation well per unit stream stage change
- Distance from the streambed to the observation well
- Width of the streambed
- Stream stage changes

Groundwater and surface water budget information for MODFLOW is included in the Well and River package data-input files. Required pumped well information for each stress period that should be defined in the conceptual model includes:

- Layer number of the model cell that contains the pumped well
- Row number of the model cell that contains the pumped well
- Column number of the model cell that contains the pumped well
- Volumetric pumped well discharge rate

A stream is divided into reaches so that each reach is completely contained in a single cell for MODFLOW. Required river information for each stress period and each river reach that should be defined in the conceptual model includes:

- Layer number of the model cell that contains the river reach
- Row number of the model cell that contains the river reach
- Column number of the model cell that contains the river reach
- Head in the river
- Riverbed hydraulic conductance
- Elevation of the bottom of the riverbed

FLOW SYSTEM EXTERNAL INFLUENCES

Analytical and numerical aquifer test analysis assumes no flow occurs in the aquifer test domain prior to the test and only one stress, the aquifer test pumped or slugged well affects groundwater levels in the aquifer test domain during the test. Compliance with these assumptions requires a horizontal groundwater level trend in the aquifer test domain prior to the start of the test and no groundwater level fluctuations during the test caused by atmospheric changes, surface water or tidal stage changes, application of heavy loads in the aquifer test domain, earth tides, earthquakes, and changes in discharge rates of nearby pumping wells. Usually, there is some flow in the aquifer test domain prior to the test, however small, and one or more stresses in addition to the aquifer test pumped well can affect groundwater levels during the aquifer test. These flow system external influences should be described in the conceptual model.

Groundwater level data collected in the aquifer test domain prior to pumping or groundwater level data collected immediately outside the aquifer test domain during pumping can be used to adjust measured time-drawdown data for the effects of any flow external influences during the aquifer test. Required flow system external influence conceptual model information includes:

- Antecedent groundwater level trend data for times prior to, during, and after the test
- Antecedent stress data such as atmospheric pressure and stream stage change records for times prior to, during, and after the test
- Flow system external influence data for times prior to, during, and after the test

CONCEPTUAL MODEL CONTENTS

Conceptual model contents for analytical aquifer test modeling software such as WTAQ differ from conceptual model contents for numerical groundwater flow modeling software such as MODFLOW. Conceptual model contents for analytical aquifer test modeling software should include:

- Map showing the locations of aquifer test pumped, slugged, and observation wells or piezometers; any interference wells; any aquifer boundaries or discontinuities with associated image wells; any stresses such as a stream with a changing stage; and the aquifer test domain radius
- Cross section showing the aquifer type, vertical dimensions of hydrostratigraphic units, and well dimensions
- Table showing pumped or slugged and observation well and piezometer dimensions such as radius, depth, screen length, and wellbore skin thickness
- Tables, graphs, and maps showing interactive calibration target data
- Tables or maps showing initial, upper, and lower hydraulic characteristic values
- Tables and graphs showing groundwater level and stress (pumping rate or initial displacement, atmospheric pressure changes, and any stream stage changes) or tidal fluctuations prior to, during, and after the pumping

Conceptual model contents for numerical groundwater flow modeling software, such as MODFLOW, should include:

- Map showing the locations of aquifer test pumped and observation wells, any interference wells, any aquifer boundaries, and the aquifer test domain grid
- Table showing grid widths and lengths
- Cross section showing layer type, layer dimensions, and well dimensions
- Table showing pumped and observation well dimensions such as radius and layer location
- Tables showing stress period, time step, and pumping rate dimensions
- Tables, graphs, and maps showing interactive calibration target data

- Tables or maps showing initial, upper, and lower layer hydraulic characteristic values
- Tables and graphs showing groundwater level and stress (atmospheric pressure and any stream stage changes) conditions prior to, during, and after the pumping test

The content of the conceptual model represents the sum total of what is known and imagined about the aquifer test domain and facilities at any particular time. The conceptual model is subject to change, especially during model calibration, until aquifer test modeling is concluded.

3 Mathematical Modeling Equation and Software Selection

The second step in aquifer test modeling is the selection of appropriate aquifer test mathematical modeling equations compatible with the previously defined conceptual model. There are two types of aquifer test mathematical modeling equations:

1. Analytical
2. Numerical

Analytical modeling equations are generally selected when a conceptual model consists of two or less layers and aquifer parameters are fairly uniform in space. Numerical modeling equations are generally selected when a conceptual model consists of more than two layers or the aquifer parameters are highly heterogeneous. It is of interest to note that analytical modeling equation solutions are often verified independently using numerical modeling equations and vice versa.

Both analytical and numerical modeling equations are based on the fundamental principles of conservation of energy, momentum, and mass. These principles and empirical laws are expressed in groundwater flow partial differential equations, which, in turn, are solved analytically or numerically subject to appropriate initial and boundary conditions and source functions to constitute aquifer test mathematical modeling equations. For details concerning principles, empirical laws, partial differential equations, and mathematical modeling equations see Hantush (1964), Remson et al. (1971), Bear (1972), Kinzelbach (1986), Bear and Veruijt (1987), Tien-Chang Lee (1999), and Cheng (2000).

ANALYTICAL MATHEMATICAL MODELING EQUATIONS

There are two types of analytical mathematical modeling equations:

1. Stehfest algorithm
2. Integral and empirical

Integral and empirical mathematical modeling equations are special cases of Stehfest algorithm mathematical modeling equations with certain parameter values assumed to be negligible. The simulation of wellbore storage and skin, well partial penetration, observation well delayed response, and delayed drainage at the water table is more difficult with integral and empirical modeling equations than with Stehfest algorithm mathematical modeling equations.

Groundwater flow partial differential equations contain a first order differential in time thereby making it possible for the equations to be solved for specified aquifer and well conditions with an integral transform called the Laplace transform. In turn, Laplace transform solutions are analytically or numerically inverted to obtain analytical aquifer test mathematical modeling equations.

Analytical inversion is based on operational calculus (contour integration) and involves complicated semidefinite integrals that are difficult or impossible to evaluate except for simple aquifer and well conditions and, in some instances, short and long elapsed time ranges. For example, analytical inversion under confined leaky with confining unit storativity or unconfined conditions is restricted to certain short and long elapsed time ranges making it difficult to obtain aquifer test mathematical modeling equations covering the entire aquifer test time.

Aquifer test modeling equations based on analytical inversion solutions (herein called the integral and empirical type) are commonly referred to as the Theis or Cooper-Jacob method (pumping test confined nonleaky aquifer), Cooper et al. method (slug test confined aquifer), Hvorslev method (slug test confined nonleaky), Hantush and Jacob method (pumping test confined leaky aquifer), Neuman method (pumping test unconfined aquifer), Bouwer and Rice method (slug test unconfined), and Springer and Gelhar method (slug test high conductivity). Methods are described in detail by Kruseman and de Ridder (1994) and Butler (1998b).

Method assumptions often include fully penetrating wells with no wellbore storage and no aquifer boundaries. In addition, the Hantush and Jacob method assumes the confining unit storativity is negligible. Several elaborate procedures have been developed to determine the best method and graph type (semilogarithmic or double logarithmic) to use with aquifer test data (American Society for Testing Materials [ASTM] guidelines). These guidelines can be downloaded at www.astm.org/cgi-bin/SoftCart.exe/index.shtml?E+mystore.

Method restrictions, simplifying assumptions, and subjective decisions can be largely avoided by using aquifer test mathematical modeling equations based on numerical inversion solutions (herein called the Stehfest algorithm). Many numerical inversion algorithms are available (Davies and Martin, 1979). Algorithms applied to the analysis of aquifer test data (Novakowski, 1990, pp. 99–107) are those developed by Stehfest (1970a, 1970b), Crump (1976), and Talbot (1979). The Stehfest algorithm is the most commonly used algorithm. The Stehfest algorithm is a polynomial approximation and can be used with available groundwater flow Laplace transform equations (Moench and Ogata, 1984, pp. 150–151).

INTEGRAL AND EMPIRICAL MATHEMATICAL MODELING EQUATIONS

The most commonly used integral aquifer test mathematical modeling equations are frequently labeled:

- Confined nonleaky pumping test — Theis (1935)
- Confined nonleaky slug test — Cooper et al. (1967)
- Confined nonleaky slug test — Hvorslev (1951)
- Confined leaky without confining unit storativity pumping test — Hantush and Jacob (1955)
- Unconfined pumping test — Neuman and Witherspoon (1972) and Neuman (1975a)
- Unconfined slug test — Bouwer and Rice (1976)
- Confined or unconfined high conductivity slug test — McElwee et al. (1992); Springer and Gelhar (1991)

Infrequently used but important integral aquifer test mathematical models are frequently labeled:

- Confined leaky pumping test with confining unit observation wells — Neuman and Witherspoon (1972)
- Induced streambed infiltration pumping test — Rorabaugh (1956) and Zlotnik and Huang (1999)

Other integral aquifer test mathematical modeling equations are described by Kruseman and de Ridder (1994) and Batu (1998).

The integral confined nonleaky pumping test mathematical modeling equation (Theis, 1935, pp. 519–524) is:

$$s = QW(u)/4\pi T \tag{3.1}$$

where

$$u = r^2S/(4Tt) \tag{3.2}$$

s is drawdown, Q is the pumped well discharge rate, T is the aquifer transmissivity, r is the distance from the pumped well, S is the aquifer storativity, and t is the elapsed time.

Major integral confined nonleaky pumping test mathematical modeling equation assumptions are:

- The aquifer is confined by overlying and underlying impermeable deposits.
- There are no aquifer boundaries or discontinuities within the cone of depression.

- The aquifer is homogeneous, isotropic, and of uniform thickness within the cone of depression.
- Prior to pumping, the piezometric surface is horizontal and piezometric pressure is constant.
- The pumping rate is constant.
- The wells penetrate and are open to the entire aquifer thickness so that flow is horizontal and not vertical in the aquifer.
- The wells have infinitesimal diameters and no wellbore storage.
- The wells have no wellbore skin.
- The pumped well has no well loss.
- During pumping, groundwater levels remain above the aquifer top.

$W(u)$ is dimensionless drawdown (well function) and u is dimensionless time. When $u \leq 0.01$ then $W(u) = -0.5772 - \ln(u) = \ln(0.562/u)$ (Hantush, 1964, p. 321). If $u < 0.25$ then $W(u) = \ln(0.78/u)$ and if $u > 1$, then $W(u) = \exp(-1.2u - 0.60)$.

Values of $W(u)$ are commonly calculated with the following polynomial approximation (Abramowitz and Stegun, 1964, p. 231):
when $0 < u \leq 1$

$$W(u) = -\ln u + a_0 + a_1 u + a_2 u^2 + a_3 u^3 + a_4 u^4 + a_5 u^5 \tag{3.3}$$

where

$$\begin{aligned} a_0 &= -0.57721566 & a_3 &= 0.05519968 \\ a_1 &= 0.99999193 & a_4 &= -0.00976004 \\ a_2 &= -0.24991055 & a_5 &= 0.00107857 \end{aligned}$$

when $1 < u < \infty$

$$W(u) = [(u^4 + a_1 u^3 + a_2 u^2 + a_3 u + a_4)/ (u^4 + b_1 u^3 + b_2 u^2 + b_3 u + b_4)]/[u\exp(u)] \tag{3.4}$$

where

$$\begin{aligned} a_1 &= 8.5733287401 & b_1 &= 9.5733223454 \\ a_2 &= 18.0590169730 & b_2 &= 25.6329561486 \\ a_3 &= 8.6347608925 & b_3 &= 21.0996530827 \\ a_4 &= 0.2677737343 & b_4 &= 3.9584969228 \end{aligned}$$

The integral confined nonleaky with fully penetrating well slug test mathematical modeling equation (Cooper et al., 1967) is:

$$H/H_0 = W(\alpha,\beta) \tag{3.5}$$

where

$$\alpha = r_w^2 S / r_c^2 \tag{3.6}$$

$$\beta = K_h bt/r_c^2 = (1/u)(r_w^2 S/4r_c^2) \tag{3.7}$$

H/H_0 is the dimensionless normalized head in the slugged well, H is the deviation of the head in the slugged well from static conditions, H_0 is the initial head change (displacement) in the slugged well, $W(\alpha,\beta)$ is the dimensionless normalized head, β is dimensionless time, r_w is the slugged well effective radius, r_c is the slugged well casing radius, S is the aquifer storativity, K_h is the aquifer horizontal hydraulic conductivity, b is the aquifer thickness, and t is the elapsed time.

Major integral confined nonleaky with fully penetrating well slug test mathematical modeling equation assumptions are:

- The aquifer is confined by overlying and underlying impervious deposits.
- There are no aquifer boundaries or discontinuities within the area of influence of the slug.
- The aquifer is homogeneous, isotropic, and of uniform thickness within the area of influence of the slug.
- Prior to the slug test, the piezometric surface is horizontal and piezometric pressure is constant.
- The slugged well penetrates and is open to the entire aquifer thickness so that flow is horizontal and not vertical in the aquifer.
- The slugged well has wellbore storage.
- The slugged well has no wellbore skin.
- During the slug test, groundwater levels remain above the aquifer top.

The Laplace transform solution for Equation 3.5 is presented by Novakowski (1989, p. 2379). The aquifer thickness can be replaced by the effective screen length in the case of a partially penetrating well (Butler, 1998b, pp. 82–87). Partially penetrating slugged well Stehfest algorithm equations were developed by Dougherty and Babu (1984, pp. 1116–1122) and Dougherty (1989, pp. 567–568).

Moench and Hsieh (1985, p. 20) present an equation for analysis of slugged well test data accounting for a skin of finite thickness. The equation assumes a confined nonleaky aquifer and a fully penetrating well. Families of type curves generated with that equation for different values of the ratio of the aquifer hydraulic conductivity to the skin hydraulic conductivity have nearly identical shapes except for very low ratios. Therefore, there is a large degree of nonuniqueness in matching test data to a family of type curves. Accurate estimates of aquifer hydraulic conductivity cannot be obtained under most circumstances, and it is not possible to tell whether there is a skin with a different hydraulic conductivity than that of the aquifer. Butler (1998b, pp. 172–176) describes the Ramey et al. method for analyzing fully penetrating slug test data that takes into account well skin.

The empirical confined nonleaky slug test mathematical modeling equation (Hvorslev, 1951) is:

$$\ln[H(t)/H_0] = -2K_h bt/[r_c^2 \ln(1/(2\Psi) + \{1 + [1/(2\Psi)]^2\}^{0.5})] \tag{3.8}$$

where

$$\Psi = (K_v / K_h)^{0.5} / (b_e / r_w) \quad (3.9)$$

when the screen bottom is above the aquifer base

$$K_h = r_c^2 \ln(1/(2\Psi) + \{1 + [1/(2\Psi)]^2\}^{0.5}) / (2b_e T_b) \quad (3.10)$$

when the screen bottom abuts the aquifer base

$$K_h = r_c^2 \ln(1/(\Psi) + \{1 + [1/(\Psi)]^2\}^{0.5}) / (2b_e T_b) \quad (3.11)$$

where K_h is the aquifer horizontal hydraulic conductivity, b is the aquifer thickness, t is the elapsed time, r_c is the well casing radius, K_v is the aquifer vertical hydraulic conductivity, b_e is the effective screen length, and T_b is the basic time lag, the time at which a normalized head of 0.368 is obtained.

Major empirical confined nonleaky slug test mathematical modeling equation assumptions are:

- The aquifer is confined by overlying and underlying impervious deposits.
- There are no aquifer boundaries or discontinuities within the area of influence of the slug.
- The aquifer is homogeneous, isotropic, and of uniform thickness within the area of influence of the slug.
- Prior to the slug test, the piezometric surface is horizontal and piezometric pressure is constant.
- The slugged well has wellbore storage.
- The slugged well has no wellbore skin.
- Groundwater levels remain above the aquifer top during the slug test.

The integral confined leaky without confining unit storativity pumping test mathematical modeling equation (Hantush and Jacob, 1955, p. 98) is:

$$s = QW(u, b_c) / 4\pi T \quad (3.12)$$

with

$$u = r^2 S / (4Tt) \quad (3.13)$$

$$b_c = r / (Tb'/K')^{0.5} \quad (3.14)$$

s is drawdown, Q is the pumped well discharge rate, T is the aquifer transmissivity, r is the distance from the pumped well to the observation well, S is the aquifer storativity, t is elapsed time, K' is the confining unit vertical hydraulic conductivity, and b' is the confining unit thickness.

Major confined leaky without confining unit storativity pumping test mathematical modeling equation assumptions are:

- The aquifer is confined by an overlying permeable confining unit and an underlying impermeable unit.
- The storativity of the confining unit is negligible.
- There is vertical leakage from the confining unit into the aquifer.
- There are no aquifer boundaries or discontinuities within the cone of depression.
- The aquifer is homogeneous, isotropic, and of uniform thickness within the cone of depression.
- Prior to pumping, the piezometric surface is horizontal and piezometric pressure is constant.
- The pumping rate is constant.
- The wells penetrate and are open to the entire aquifer thickness so that flow is horizontal and not vertical in the aquifer.
- The wells have infinitesimal diameters and no wellbore storage.
- The wells have no wellbore skin.
- The pumped well has no well loss.
- During pumping, groundwater levels remain above the aquifer top.

Values of $W(u,b_c)$ are commonly calculated with the following equations (Wilson and Miller, 1978, p. 505):
when $r/b_c > 2$

$$W(u,b_c) = [\pi/(2r/b_c)]^{0.5}\exp(-r/b_c)\mathrm{erfc}[-(r/b_c - 2u)/(2u^{0.5})] \quad (3.15)$$

Values of erfc(x) are commonly calculated with the following approximations presented by Abramowitz and Stegun (1964):

$$\mathrm{erfc}(x) = 1/[1 + a_1(x) + a_2(x)^2 + \ldots + a_6(x)^6]^{16} \quad (3.16)$$

$$\mathrm{erfc}(-x) = 1 + \mathrm{erf}(x) \quad (3.17)$$

where

$$\mathrm{erf}(x) = 1 - 1/[1 + a_1(x) + a_2(x)^2 + \ldots + a_6(x)^6]^{16} \quad (3.18)$$

$$\mathrm{erf}(-x) = -\mathrm{erf}(x) \quad (3.19)$$

and

$a_1 = 0.0705230784$ $a_4 = 0.0001520143$
$a_2 = 0.0422820123$ $a_5 = 0.0002765672$
$a_3 = 0.0092705272$ $a_6 = 0.0000430638$

when $r/b_c < 2$ and $r/b_c > 0$ and $(r/b_c)^2/(4u) > 5$

$$W(u,b_c) = 2K_0(r/b_c) \tag{3.20}$$

when $r/b_c \leq 2$ and $(r/b_c)^2/(4u) \leq 5$

$$W(u,b_c) = E_1(u) + \sum_{m=1}^{\infty}(1/m!) - [(r/b_c)^2/(4u)]^m E_{m+1}(u) \tag{3.21}$$

with

$$E_1(u) = W(u) \tag{3.22}$$

$$E_{m+1}(u) = (1/m)[\exp(u) - uE_m(u)] \tag{3.23}$$

$$m! = m(2\pi/m)^{0.5}m^m\exp(y) \tag{3.24}$$

with

$$y = 1/(12m) - 1/(360m^3) - m \tag{3.25}$$

Values of $K_0(r/b_c)$ are commonly calculated with the following polynomial approximations presented by Abramowitz and Stegun (1964):

when $0 < r/b_c \leq 2$

$$\begin{aligned} K_0(r/b_c) = {} & -\ln[(r/b_c)/2]I_0(r/b_c) - 0.57721566 + 0.42278420[(r/b_c)/2]^2 \\ & + 0.23069756[(r/b_c)/2]^4 + 0.03488590[(r/b_c)/2]^6 \\ & + 0.0026298[(r/b_c)/2]^8 + 0.00010750[(r/b_c)/2]^{10} \\ & + 0.00000740[(r/b_c)/2]^{12} \end{aligned} \tag{3.26}$$

with

$$\begin{aligned} I_0(r/b_c) = {} & 1 + 3.5156229[(r/b_c)/3.75]^2 + 3.0899424[(r/b_c)/3.75]^4 \\ & + 1.2067492[(r/b_c)/3.75]^6 + 0.2659732[(r/b_c)/3.75]^8 \\ & + 0.0360768[(r/b_c)/3.75]^{10} + 0.0045813[(r/b_c)/3.75]^{12} \end{aligned} \tag{3.27}$$

Equation 3.27 is valid when $-3.75 \leq r/b_c \leq 3.75$.

When $2 < r/b_c < \infty$

$$\begin{aligned} K_0(r/b_c) = {} & \{1.25331414 - 0.07832358[2/(r/b_c)] + 0.02189568[2/(r/b_c)]^2 \\ & - 0.01062446[2/(r/b_c)]^3 + 0.00587872[2/(r/b_c)]^4 - 0.00251540[2/(r/b_c)]^5 \\ & + 0.00053208[2/(r/b_c)]^6\}/[(r/b_c)^{0.5}\exp(r/b_c)] \end{aligned} \tag{3.28}$$

The integral unconfined aquifer pumping test mathematical modeling equation (Neuman, 1975a) is:

$$s = QW(u_a, u_b, \beta, \sigma)/4\pi T \quad (3.29)$$

where

$$u_a = r^2 S/(4Tt) \quad (3.30)$$

$$u_b = r^2 S_y/(4Tt) \quad (3.31)$$

$$\beta = (r^2 K_v)/(b^2 K_h) \quad (3.32)$$

$$\sigma = S/S_Y \quad (3.33)$$

$$u_b = \sigma u_a \quad (3.34)$$

s is drawdown, Q is the pumped well discharge rate, T is the aquifer transmissivity, r is the distance from the pumped well to the observation well, S is the aquifer storativity, t is elapsed time, S_Y is the aquifer specific yield, K_v is the aquifer vertical hydraulic conductivity, K_h is the aquifer horizontal hydraulic conductivity, and b is the aquifer thickness.

Major integral unconfined aquifer pumping test mathematical modeling equation assumptions are:

- The aquifer is unconfined at the top and is underlain by impermeable deposits.
- Delayed gravity of upper portions of the aquifer occurs during the pumping period.
- There is instantaneous drainage at the water table.
- The portion of the aquifer dewatered during the pumping period is negligible.
- There are no aquifer boundaries or discontinuities within the cone of depression.
- The aquifer is homogeneous, isotropic, and of uniform thickness within the cone of depression.
- Prior to pumping, the piezometric surface is horizontal and piezometric pressure is constant.
- The pumping rate is constant.
- The wells penetrate and are open to the entire aquifer thickness so that flow is horizontal and not vertical in the aquifer.
- The wells have infinitesimal diameters and no wellbore storage.
- The wells have no wellbore skin.
- The pumped well has no well loss.

The empirical unconfined aquifer slug test mathematical modeling equation (Bouwer and Rice, 1976; Zlotnik, 1994) is:

$$\ln[(H(t)/H_0)] = -\{2K_h b_{sc} t/[r_c^2 \ln(R_c/r_w^*)]\} \tag{3.35}$$

where

$$r_w^* = (K_v/K_h)^{0.5} \tag{3.36}$$

$$\ln(R_c/r_w^*) = \{1.1/\ln[(d + b_{sc})/r_w^*] + D\}^{-1} \tag{3.37}$$

When the well terminates above the aquifer base

$$D = (A + B\{\ln[b - (d + b_{sc})]/r_w^*\})/(b/r_w^*) \tag{3.38}$$

When the term $\{\ln[b - (d + b)]/r_w^*\}$ is greater than 6.0 then the term should be 6.0. When the well terminates at the aquifer base (fully penetrating well)

$$D = C/(b_{sc}/r_w^*) \tag{3.39}$$

K_h is the aquifer horizontal hydraulic conductivity, r_c is the well casing radius, b is the aquifer thickness, t is the elapsed time, K_v is aquifer vertical hydraulic conductivity, d is the z position of the end of the screen, b_{sc} is the screen length.

The empirical coefficients A, B, and C can be calculated with the following expressions (Boak, 1991; Van Rooy, 1988):

$$\begin{aligned} A = {} & 1.4720 + 3.537 \times 10^{-2}(b_{sc}/r_w^*) - 8.148 \times 10^{-5}(b_{sc}/r_w^*)^2 \\ & + 1.028 \times 10^{-7}(b_{sc}/r_w^*)^3 - 6.484 \times 10^{11}(b_{sc}/r_w^*)^4 \\ & + 1.573 \times 10^{-14}(b_{sc}/r_w^*)^5 \end{aligned} \tag{3.40}$$

$$\begin{aligned} B = {} & 0.2372 + 5.151 \times 10^{-3}(b_{sc}/r_w^*) - 2.682 \times 10^{-6}(b_{sc}/r_w^*)^2 \\ & - 3.491 \times 10^{-10}(b_{sc}/r_w^*)^3 + 4.738 \times 10^{-13}(b_{sc}/r_w^*)^4 \end{aligned} \tag{3.41}$$

$$\begin{aligned} C = {} & 0.7290 + 3.993 \times 10^{-2}(b_{sc}/r_w^*) - 5.743 \times 10^{-5}(b_{sc}/r_w^*)^2 \\ & + 3.858 \times 10^{-8}(b_{sc}/r_w^*)^3 - 9.659 \times 10^{-12}(b_{sc}/r_w^*)^4 \end{aligned} \tag{3.42}$$

The logarithm of the normalized head data is plotted vs. the elapsed time; a straight line is fitted to the data over the time-normalized head interval 0.20 to 0.30; the slope of the line is calculated by estimating the time (T_0) at which a normalized head of 0.368 is obtained, and the aquifer horizontal hydraulic conductivity (K_h) is estimated with the following equation:

$$K_h = r_c^2[\ln(R_c/r_w^*)]/(2b_{sc}T_0) \tag{3.43}$$

where r_c is the well casing radius and b_{sc} is the screen length.

Equation 3.43 is valid for wells screened below the water table. For wells screened across the water table, a straight line is fitted to the second segment of the time-normalized head graph and the following equation is used to estimate the term r_c (Bouwer, 1989):

$$r_c = [r_{nc}^2 + n(r_{wfp}^2 - r_{nc}^2)]^{0.5} \tag{3.44}$$

where r_{nc} is the nominal radius of the well screen, r_{wfp} is the outer radius of the filter pack, and n is the drainable porosity of the filter pack.

Major integral unconfined aquifer slug test mathematical modeling equation assumptions are:

- The aquifer is unconfined at the top and is underlain by an impermeable unit.
- There is instantaneous gravity drainage of a small upper portion of the aquifer.
- The portion of the aquifer dewatered during the slug test is negligible in comparison to the original aquifer thickness.
- Delayed drainage at the water table is negligible.
- The effects of elastic storage mechanisms are negligible.
- There are no aquifer boundaries or discontinuities within the area of influence of the slug.
- The aquifer is homogeneous, isotropic or anisotropic, and of uniform thickness within the area of influence of the slug.
- Prior to slug test, the water table is horizontal and groundwater levels are constant.
- The slugged well has wellbore storage.
- The slugged well has no wellbore skin.

Slug well test response data are oscillatory in nature in aquifers of very high hydraulic conductivity (Bredehoeft et al., 1966; Van der Kamp, 1976). The *equation for analyzing slugged well test data for formations of very high hydraulic conductivity* is based on solutions to the following equation (Butler, 1998b, p. 155):

$$d^2w_d/dt_d^2 + C_d dw_d/dt_d + w_d = 0 \tag{3.45}$$

where

$$w_d = w/H_0 \tag{3.46}$$

$$t_d = (g/L_c)^{0.5}t \tag{3.47}$$

$$L_c = g/[\omega^2 + (C_v/2)^2] \tag{3.48}$$

$$\omega = 2\pi/(t_{n+1} - t_n) \tag{3.49}$$

$$C_v = 21n(W_n/W_{n+1})/(t_{n+1} - t_n) \tag{3.50}$$

$$C_d = C_v/(g/L_c)^{0.5} \tag{3.51}$$

for confined nonleaky aquifer conditions (McElwee et al., 1992):

$$K_h = (g/L_c)^{0.5}[r_c^2 \ln(1/(2\Psi) + \{1 + [1/(2\Psi)]^2\}^{1/2})/(2\ C_d b_{sc})] \tag{3.52}$$

where

$$\Psi = (K_v/K_h)^{0.5}/(b_e/r_w) \tag{3.53}$$

for unconfined aquifer conditions (Springer and Gelhar, 1991):

$$K_h = (g/L_c)^{0.5}[r_c^2 \ln(R_c/r_w^*)/(2C_d b_{sc})] \tag{3.54}$$

$\ln[(R_c/r_w^*)]$ is patterned after a like term in the Bouwer and Rice method, g is the acceleration due to gravity (9.754 m/sec/sec or 32 ft/sec/sec), w is the deviation of water level from static level in the slugged well, L_c is the effective column length, C_d is the dimensionless damping parameter, C_v is the damping parameter, ω is the frequency parameter, t_n is the time of the nth peak or trough in the slugged well data, b_e is the effective screen length, b_{sc} is the screen length, r_w is the well effective radius, K_v is the aquifer vertical hydraulic conductivity, K_h is the aquifer horizontal hydraulic conductivity, H_0 is the initial head change (displacement) in the slugged well, and W_n is the w value at the nth peak or trough in the slugged well data. The damping and frequency parameters are estimated from subsequent peaks or troughs in the slugged well data.

Major slugged well test data for formations of very high hydraulic conductivity mathematical modeling equation assumptions are:

- The aquifer has a high conductivity and is either confined by overlying and underlying impermeable units or unconfined at the top and underlain by impervious units.
- Slug impacts are similar to the behavior of a damped spring.
- There are no aquifer boundaries or discontinuities within the area of influence of the slug.
- The aquifer is homogeneous, isotropic or anisotropic, and of uniform thickness within the area of influence of the slug.
- Prior to the slug test, the piezometric surface or water table is horizontal and piezometric pressure or the groundwater level is constant.

- The slugged well has wellbore storage.
- The slugged well has no wellbore skin.

Sometimes, a confined leaky aquifer system has a very low confining unit vertical hydraulic conductivity. Under this condition, it is often impossible to determine confining unit parameter values using data for wells in the aquifer because the effects of leakance are too small during a normal pumping test period to be analyzed with any reasonable degree of accuracy. However, it is possible to determine the aquifer parameter values using the aquifer well data.

When the confining unit has a very low vertical hydraulic conductivity, a piezometer is constructed in the confining unit a few feet above the aquifer top and at the same location as one of the aquifer observation wells. The nested aquifer observation well and the confining unit piezometer must be close to the pumped well and water levels in the aquifer observation well and the confining unit piezometer must be measured at the same elapsed time.

The *integral confined leaky pumping test with confining unit observation wells* (Neuman and Witherspoon, 1972) mathematical modeling equation is:

$$s_c = QW(u,u_c)/(4\pi T) \tag{3.55}$$

where

$$u_c = z^2S'/(4K'b't) \tag{3.56}$$

s_c is the drawdown in the confining unit, Q is the pumped well discharge rate, T is the aquifer transmissivity, z is the vertical distance from the aquifer top to the base of the confining unit piezometer, S' is the confining unit storativity, K' is the confining unit vertical hydraulic conductivity, b' is the confining unit thickness, and t is the elapsed time.

Assuming that an aquifer observation well and a confining unit piezometer are located at the same short radial distance (< 300 ft) from the pumped well and drawdowns in the aquifer observation well and the confining unit piezometer are measured at the same elapsed time, the ratio of the drawdown in the confining unit (s_c) and the drawdown (s) in the aquifer is:

$$s_c/s = W(u,u_c)/W(u) \tag{3.57}$$

Values and curves of $W(u,u_c)/W(u)$ versus $1/u_c$ for different values of u are given by Kruseman and de Ridder (1994, pp. 94–95). A value of $1/u_c$ is interpreted from the values or curves based on Equation 3.55, the measured s_c/s ratio, and a previously determined value of u for the aquifer. The ratio S/K is calculated with Equation 3.56.

The *integral induced streambed infiltration pumping test* (Rorabaugh, 1956; Zlotnik and Huang, 1999) mathematical modeling equations are commonly used to estimate aquifer parameter values and the streambed vertical hydraulic conductivity. During an induced infiltration pumping test, a well near a stream is pumped and drawdowns are measured in several nearby observation wells on a ray through the pumped well and parallel to the stream. Prior to the test, water level changes in an observation well near the stream due to stream stage changes are measured. This data is used to determine the streambed leakance.

Aquifer transmissivity is calculated with distance-drawdown data for observation wells on a ray through the pumped well and parallel to the stream measured at the end of the test after groundwater levels stabilize. These observation wells are approximately equidistant from the recharging image well simulating the stream. Thus the effects of induced infiltration on water levels in these wells are approximately equal and the hydraulic gradient of the cone of depression near the production well and parallel to the stream is not distorted to any appreciable degree. A plot of drawdown in the observation wells parallel to the stream vs. the logarithm of the distances between the pumped and observation wells yields a straight line. The slope of the straight line and the pumping rate are inserted in the following equation to calculate aquifer transmissivity (Cooper and Jacob, 1946, pp. 526–534):

$$T = 2.3Q/(2\pi\Delta s) \tag{3.58}$$

where T is the aquifer transmissivity, Q is the pumped well discharge rate, and Δs is the drawdown per logarithmic cycle (slope of the straight line).

The distance between the pumped well the recharging image well simulating the stream is calculated with the following equation (see Rorabaugh, 1956, pp. 101–169):

$$2a = \exp(c)r \tag{3.59}$$

where

$$c = 2\pi Ts/Q \tag{3.60}$$

T is the aquifer transmissivity, s is drawdown, Q is the pumped well discharge rate, a is the distance between the pumped well and the effective line of recharge, and r is the distance between a particular observation well and the pumped well.

Several values of aquifer specific yield are assumed and observation well drawdowns for each assumed value are calculated with the following equation (Ferris et al., 1962, pp. 144–166):

$$s_o = [Q/(4\pi T)]W_b(u) \tag{3.61}$$

where

$$W_b(u) = W(u) - W(u_i) \tag{3.62}$$

$$u = r^2 S_Y/(4Tt) \quad (3.63)$$

$$u_i = r_i^2 S_Y/(4Tt) \quad (3.64)$$

s_o is the calculated drawdown in an observation well with the assumed aquifer specific yield S_Y and previously calculated aquifer transmissivity T, Q is the pumped well discharge rate, r is the distance between the pumped well and an observation well, r_i is the distance between an observation well and the recharge image well simulating the stream, and t is the elapsed time.

Calculated values of drawdown are compared with measured drawdowns and that specific yield used to calculate drawdowns equal to the measured drawdowns is assigned to the aquifer.

The streambed leakance is estimated with the previously calculated values of aquifer transmissivity and specific yield, measured water level changes in an observation well near the stream due to stream stage changes, and the following equations (Zlotnik and Huang, 1999, pp. 599–605):

$$s_{sc} = \mathrm{erfc}[(x' - 1)/(2t'^{0.5})] - \exp[\xi(x' - 1) + t'\xi^2]$$
$$\mathrm{erfc}[(x' - 1)/(2t'^{0.5}) + \xi\, t'^{0.5}] \quad (3.65)$$

where

$$x' = x/w \quad (3.66)$$

$$t' = Tt/(S_Y w^2) \quad (3.67)$$

$$\xi = \gamma^{0.5} \tanh \gamma^{0.5} \quad (3.68)$$

$$\gamma = K_s' w^2/(b_s' T) \quad (3.69)$$

s_{sc} is the water level change in an observation well per unit stream stage change, T is the aquifer transmissivity, S_Y is the aquifer specific yield, x is the distance from the streambed center to an observation well, w is the half-width of the streambed, t is the elapsed time after a stream stage change began, K_s' is the streambed vertical hydraulic conductivity, b_s' is the streambed thickness, and K_s'/b_s' is the streambed leakance. Materials beneath the streambed are assumed to be saturated and any seepage of water through unsaturated materials beneath the streambed is assumed to be negligible. If materials beneath the streambed are unsaturated, calculated streambed leakance will be less than actual leakance because the influence of negative pressure heads in unsaturated materials is ignored. In many cases, inaccuracies in estimating streambed leakance may overshadow errors due to ignoring unsaturated conditions (see Peterson, 1989, pp. 899–927).

Values of s_{sc} for selected values of streambed leakance are calculated and compared with the measured water level changes per unit stream stage change. The streambed leakance that results in an acceptable match of calculated and measured water level changes per unit stream stage change is assigned to the stream.

Wellbore Storage Effects

Some integral aquifer test modeling equations assume pumped and observation wellbore storage are negligible. If the conceptual model wells have wellbore storage, dimensionless time-drawdown values calculated with the integral aquifer test modeling equations should be adjusted for wellbore storage effects before they are used to calculate drawdown values. Otherwise, the ranges of pumping test data that are not likely to be affected appreciably by wellbore storage are subjectively selected (filtered) for analysis. Wellbore storage adjustments are the differences between dimensionless time drawdown with and without wellbore storage.

In general, pumped wellbore storage adjustments tend to be negligible except for the first few minutes of a pumping test with moderate to high (> 1000 ft²/day) transmissivities and small (< 0.5 ft) well radii as illustrated in Figure 3.1. Appreciable pumped wellbore storage adjustments are required during the first several hours or even days of the pumping test with lower transmissivities or large well radii or small storativities. Adjustments for observation wellbore storage increase as the distance from the production well and storativity decreases and the production or observation well radius increases.

Adjustments for observation wellbore storage are required close to the pumped well (within tens of feet) during early elapsed times under most aquifer

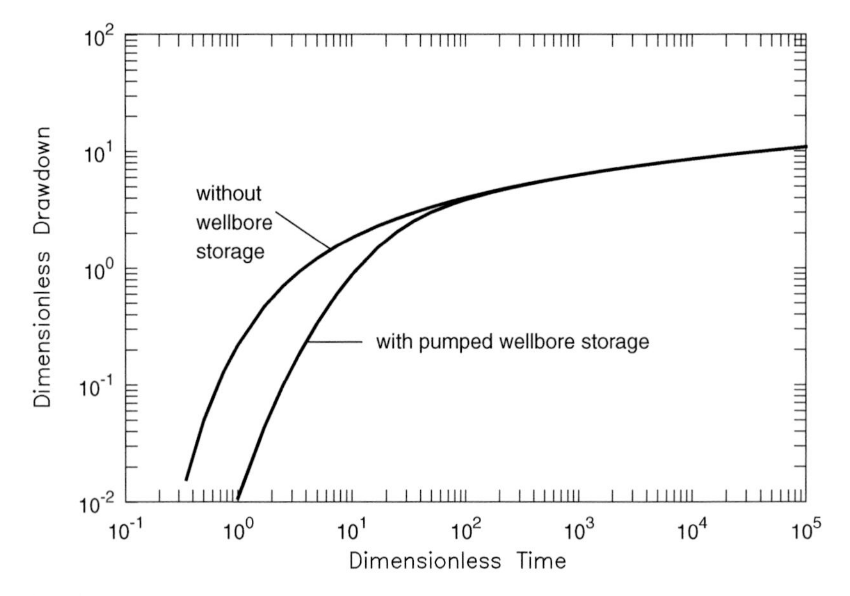

FIGURE 3.1 Graph showing wellbore storage effects on pumping test type curve values.

conditions. Both pumped and observation wellbore storage adjustments decrease with time (Fenske, 1977). Commonly, with storativity in the 1×10^4 range and at distances exceeding several tens of feet beyond the production well, the effect of observation wellbore is negligible and adjustments are not required (Fenske, 1977; Tongpenyai and Raghavan, 1981).

The effects of pumped and observation wellbore storage are dependent on the relative quantities of discharge derived from well storage and aquifer storage. During early elapsed times when change in drawdown is rapid, a large portion of the discharge is derived from well storage. During later elapsed times when change in drawdown is very slow, a large portion of the discharge is derived from aquifer storage. Observation wells contain a significant quantity of stored water. With the start of pumping, rapid changes in hydraulic head in the aquifer may not be accurately reflected by measurements in observation wells because of the finite time it takes to dissipate stored water and reach equilibrium with the hydraulic head in the aquifer.

The ratio of the discharge derived from aquifer storage (Q_{aw}) and the discharge rate (Q) for infinite confined nonleaky aquifers with fully penetrating wells is as follows (Fenske, 1977, p. 90):
without pumped and observation wellbore storage

$$Q_{aw}/Q = 1 \tag{3.70}$$

with pumped wellbore storage

$$\begin{aligned} Q_{ap}/Q = 1 - Q_w/Q = 1 - 1/\alpha\ \{[e^{-n} + W(n)]/[(ne^{n})^{-1} - (1 - 1/\alpha)W(n)] \\ - W(n)[e^{-n}(1/n + 1/\alpha)]/[(ne^{-n})^{-1}(1 - 1/\alpha)W(n)]^2\} \end{aligned} \tag{3.71}$$

with pumped and observation wellbore storage

$$\begin{aligned} Q_{apo}/Q = 1 - Q_w/Q = 1 - 1/\alpha\{[e^{-n} + W(n)]/[(ne^{n})^{-1} - (1 - 1/\alpha)W(n) \\ + (1/\beta_m)W(m)] - W(n)[e^{-n}\ (1/n + 1/\alpha) + (1/\beta_m)e^{-m}]/[(ne^{n})^{-1} \\ - 1(1 - 1/\alpha)W(n) + (1/\beta_m)W(m)]^2\} \end{aligned} \tag{3.72}$$

where

$$n = Sr_w^2/(4Tt) \tag{3.73}$$

$$m = Sr^2/(4Tt) \tag{3.74}$$

$$\alpha = [r_w^2/(r_c^2 - r_d^2)]S \tag{3.75}$$

$$\beta_m = (r_w^2/r_o^2)S/(1 - S) \tag{3.76}$$

Q_{aw} is the discharge from the aquifer without pumped and observation wellbore storage, Q_{ap} is the discharge from the aquifer with pumped wellbore storage, Q_{apo} is the discharge from the aquifer with pumped and observation wellbore storage Q is the total discharge, S is the aquifer storativity, r_w is the pumped well effective radius, T is the aquifer transmissivity, t is the elapsed time, r is the distance from the pumped well to the observation well, r_c is the pumped well casing radius, r_p is the pump pipe radius, and r_o is the observation well casing radius.

Values of dimensionless drawdown $W(n)$ and $W(m)$ are commonly calculated with a polynomial approximation previously described for $W(u)$ (Abramowitz and Stegun, 1964, p. 231).

Both dimensionless time and dimensionless drawdown values are affected by wellbore storage. Dimensionless time values for early elapsed times are shifted to the right (dimensionless time values increase) due to wellbore storage effects and dimensionless drawdown values for early elapsed times are shifted downward (dimensionless drawdown values decrease) due to wellbore storage effects. Dimensionless time adjustments for wellbore storage effects can be calculated with the following equations (Fenske, 1977, p. 89):

The dimensionless time adjustment for pumped wellbore storage $1/u_{adp}$ is:

$$1/u_{adp} = 1/u_{pw} - 1/u \tag{3.77}$$

The dimensionless time adjustment for observation wellbore storage $1/u_{ado}$ is:

$$1/u_{ado} = 1/u_{pow} - 1/u_{pw} \tag{3.78}$$

The dimensionless time adjustment for both pumped and observation wellbore storage $1/u_{adpo}$ is:

$$1/u_{adpo} = 1/u_{pow} - 1/u \tag{3.79}$$

where:

dimensionless time without wellbore storage

$$1/u = 4Tt/(r^2S) \tag{3.80}$$

dimensionless time with pumped wellbore storage

$$1/u_{pw} = (Q_{ap}/Q)[(ne^n)^{-1} - (1 - 1/\alpha)W(n)] \tag{3.81}$$

dimensionless time with both pumped and observation wellbore storage

$$1/u_{pow} = (Q_{apo}/Q)[(ne^n)^{-1} - (1 - 1/\alpha)W(n) + (1/\beta_m)W(m)] \tag{3.82}$$

Dimensionless drawdown adjustments for wellbore storage effects can be calculated with the following equations (Fenske, 1977, p. 89):

The dimensionless drawdown adjustment for pumped wellbore storage $W(u)_{adp}$ is:

$$W(u)_{adp} = W(u)_{pw} - W(u) \tag{3.83}$$

The dimensionless drawdown adjustment for observation wellbore storage $W(u)_{ado}$ is:

$$W(u)_{ado} = W(u)_{pow} - W(u)_{pw} \tag{3.84}$$

The dimensionless drawdown adjustment for both pumped and observation wellbore storage $W(u)_{adpo}$ is:

$$W(u)_{adpo} = W(u)_{pow} - W(u) \tag{3.85}$$

where:

dimensionless drawdown without wellbore storage

$$W(u) \tag{3.86}$$

dimensionless drawdown with pumped wellbore storage

$$W(u)_{adp} = (Q_{ap}/Q)W(u) \tag{3.87}$$

dimensionless drawdown with pumped and observation wellbore storage

$$W(u)_{apo} = (Q_{apo}/Q)W(u) \tag{3.88}$$

Major wellbore storage adjustment mathematical modeling equation assumptions are:

- The aquifer is confined by overlying and underlying impermeable units.
- There are no aquifer boundaries or discontinuities within the cone of depression.
- The aquifer is homogeneous, isotropic, and of uniform thickness within the cone of depression.
- Prior to pumping, the piezometric surface is horizontal and piezometric pressure is constant.
- The pumping rate is constant.
- The wells penetrate and are open to the entire aquifer thickness so that flow is horizontal and not vertical in the aquifer.
- The wells have infinitesimal diameters and no wellbore storage.
- The wells have no wellbore skin.
- The pumped well has no well loss.
- During pumping, groundwater levels remain above the aquifer top.

Although the dimensionless time and dimensionless drawdown adjustments strictly apply to confined nonleaky aquifers, they can be applied to other aquifer conditions with little error because the adjustments are usually of significance only during early elapsed times when confined nonleaky conditions prevail under most aquifer conditions.

Well Partial Penetration Effects

Some integral aquifer test mathematical modeling equations assume fully penetrating wells. If the conceptual model wells partially penetrate the aquifer, dimensionless time drawdowns should be adjusted for the effects of partially penetrating wells values, otherwise aquifer test data that are not likely to be affected appreciably by partial penetration effects are subjectively selected for analysis. Partially penetrating pumped wells induce vertical components of flow that are assumed to be negligible in some aquifer test modeling equations. Well partial penetration adjustments are strongest at the pumped well face and decrease with increasing distance from the pumped well.

Well partial penetration adjustments may be either negative or positive depending on well geometry as illustrated in Figure 3.2. For example, if the pumped and observation wells are both open in either the top or bottom portion of the aquifer, the measured drawdown in the observation well is greater than it would be with fully penetrating wells. If the pumped well is open to the top of the aquifer and the observation well is open to the bottom of the aquifer, or vice versa, the measured drawdown in the observation well is less than it would be with fully penetrating wells.

The distance beyond which well partial penetration adjustments are negligible is defined by the following equation (Hantush, 1964, p. 351):

$$r_{pp} = 2b(K_h/K_v)^{0.5} \tag{3.89}$$

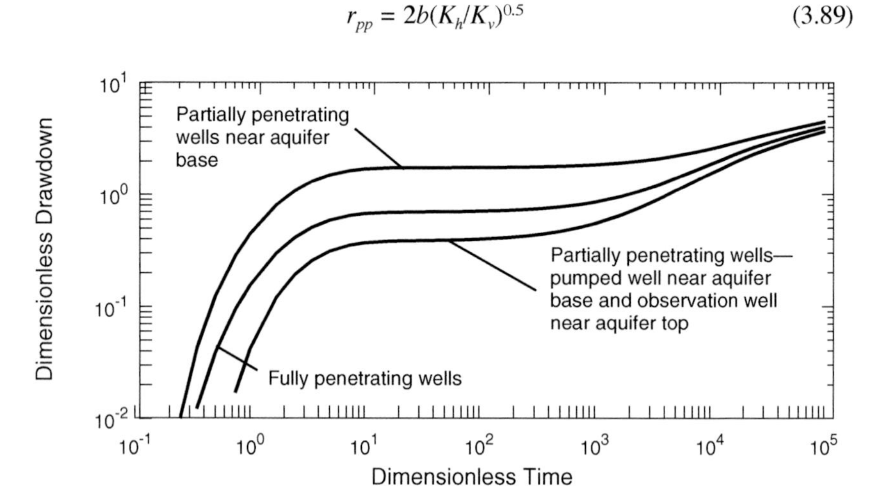

FIGURE 3.2 Graph showing well partial penetration effects on pumping test type curve values.

r_{pp} is the distance beyond which the effects of well partial penetration are negligible, b is the aquifer thickness, K_h is the aquifer horizontal hydraulic conductivity, and K_v is the aquifer vertical hydraulic conductivity.

Drawdown during the entire elapsed time in confined aquifers or during early and late elapsed times, but not intermediate elapsed times, in unconfined aquifers due to the effects of well partial penetration can be calculated with the following equations (Hantush, 1961, pp. 85 and 90; Reed, 1980, pp. 8–10):

$$s_{pp} = QW_{pp}(u\ \ldots)/(4\pi T) \tag{3.90}$$

with

$$W_{pp}(u\ \ldots) = 2b^2/[\pi^2(L - D)(L' - D')] \sum_{n=1}^{\infty} (1/n^2)[\sin(n\pi L/b) - \sin(n\pi D/b)]$$
$$[\sin(n\pi L'/b) - \sin(n\pi D'/b)]W(u,b_{pp}) \tag{3.91}$$

where

$$b_{pp} = (K_v/K_h)^{0.5} n\pi r/b \tag{3.92}$$

s_{pp} is the drawdown due to the effects of well partial penetration, T is the aquifer transmissivity, Q is the pumped well discharge rate, b is the aquifer thickness, L is the vertical distance from the aquifer top to the bottom of the pumped well screen, D is the vertical distance from the aquifer top to the top of the pumped well screen, L' is the vertical distance from the aquifer top to the bottom of the observation well screen, D' is the vertical distance from the aquifer top to the top of the observation well screen.

Well partial penetration adjustments increase during early elapsed times. For larger times $t > b^2S/[2(K_v/K_h)T]$ or $t > bS/(2K_v)$, well partial penetration adjustments gradually level off, become constant in time, and are equal to $2K_0(b_{pp})$.

Dimensionless drawdown values for fully penetrating conditions can be converted to dimensionless drawdown values for partially penetrating wells using the following equation:

$$W_p(u\ldots) = W(u\ldots) + W_{pp}(u\ldots) \tag{3.93}$$

where $W_p(u\ldots)$ is the dimensionless drawdown with well partial penetration effects, $W(u\ldots)$ is the dimensionless drawdown without the effects of well partial penetration, and $W_{pp}(u\ldots)$ is the dimensionless drawdown due to the effects of well partial penetration.

Partial penetration dimensionless drawdown values for the pumped well are calculated by substituting r_w for r and $L' = (L + D)/2$ and $D' = L'\ [0.1(L' - D)]$ (see Hantush, 1964, p. 352).

Major well partial penetration mathematical modeling equation assumptions are:

- The aquifer is confined by overlying and underlying impermeable units.
- There are no aquifer boundaries or discontinuities within the cone of depression.
- The aquifer is homogeneous, isotropic, and of uniform thickness within the cone of depression.
- Prior to pumping, the piezometric surface is horizontal and piezometric pressure is constant.
- The pumping rate is constant.
- The wells penetrate and are open to the entire aquifer thickness so that flow is horizontal and not vertical in the aquifer.
- The wells have infinitesimal diameters and no wellbore storage.
- The wells have no wellbore skin.
- The pumped well has no well loss.
- During pumping, groundwater levels remain above the aquifer top.

Pumped Well Conditions

Both integral and Stehfest algorithm aquifer test modeling equations assume that nonlinear well losses in the pumped well are negligible. If the conceptual model pumped well has well loss, drawdown in the pumped well should be adjusted for the effects of nonlinear well loss before analyzing data for the pumped well. Nonlinear well losses occur inside the well screen, in the suction pipe, and in the zone adjacent to the well where the flow is turbulent.

Well loss can be estimated with the following equation (Jacob, 1947):

$$s_{wL} = CQ^2 \tag{3.94}$$

where s_{wL} is the component of drawdown in the pumped well due to well loss, C is the well loss constant, and Q is the pumped well discharge rate.

Typical well loss constants are (Walton, 1962, p. 27): negligible well loss — 0 sec^2/ft^5, low well loss — 1 sec^2/ft^5, moderate well loss — 5 sec^2/ft^5, and severe well loss — 20 sec^2/ft^5. The well loss constant is usually estimated by conducting a step-drawdown test. The well is pumped at three or more constant fractions of full capacity for periods of one hour and drawdowns are measured during each period. Assuming that the well is stable and well loss is equal to CQ^2, the well loss constant C can be estimated with the following equation (Jacob, 1947):

$$C = (\Delta s_i/\Delta Q_i - \Delta s_{i-1}/\Delta Q_{i-1})/(\Delta Q_i + \Delta Q_{i-1}) \tag{3.95}$$

where Δs_i is the increment of drawdown at the end of pumping period i due to the increment of discharge ΔQ_i during pumping period i.

For Step 1 and Step 2:

$$C = (\Delta s_2/\Delta Q_2 - \Delta s_1/\Delta Q_1)/(\Delta Q_1 + \Delta Q_2) \quad (3.96)$$

For Step 2 and Step 3:

$$C = (\Delta s_3/\Delta Q_3 - \Delta s_2/\Delta Q_2)/(\Delta Q_2 + \Delta Q_3) \quad (3.97)$$

If the well is unstable, C cannot be calculated with data for Step 2 and Step 3 because the calculations yield a negative C. In this case, data for Step 1 and Step 2 are combined and C is estimated with the following equation (Jacob, 1947; Walton, 1991, p. 165):

For Step 1 plus Step 2 and Step 3:

$$C_{1+2 \text{ and } 3} = (\Delta s_3/\Delta Q_3 - \Delta s_{1+2}/\Delta Q_{1+2})/(\Delta Q_{1+2} + \Delta Q_3) \quad (3.98)$$

where Δs_3 is the increment of drawdown during the third pumping period with an increment of discharge rate ΔQ_3, Δs_{1+2} is the increment of drawdown during the first pumping period plus the increment of drawdown during the second pumping period, and ΔQ_{1+2} is the increment of discharge rate during the first pumping period plus the increment of discharge rate during the second pumping period.

There are more sophisticated well loss equations that may or may not improve the precision of estimates (Rorabaugh, 1953; Lennex, 1966; Hantush, 1964; Bierschenk, 1963; Eden and Hazel, 1973).

Aquifer Boundary Effects

Integral aquifer test modeling equations assume that the aquifer is infinite in areal extent. If the conceptual model aquifer is finite, drawdown should be adjusted for boundary effects as illustrated in Figure 3.3, otherwise, ranges of data not likely to be affected by boundaries must be subjectively selected for analysis.

The existence of hydrogeologic boundaries (full or partial barrier or recharge) can limit the continuity of an aquifer in one or more directions. Partial barrier or recharge boundaries are called discontinuities. Pumping test data will show the impacts of a full boundary when transmissivity in the immediate vicinity of the pumping well is ten times greater than or one tenth less than the transmissivity at some distance from the pumping well (Fenske, 1984, pp. 131–132). Adjustments for boundary effects are made with the image well theory to be explained later.

Noordbergum Confined Leaky Aquifer Effect

When water is pumped from a confined leaky aquifer, the head in the overlying or underlying confining unit can rise. This effect (called the Noordbergum effect) is attributed to three-dimensional deformation of the aquifer and confining unit when pumping starts and is not considered in integral and Stehfest algorithm

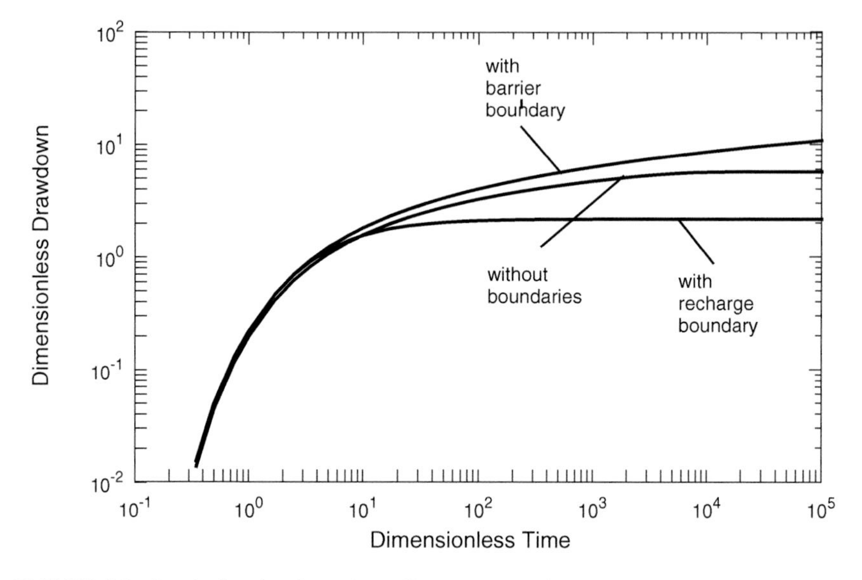

FIGURE 3.3 Graph showing boundary effects on pumping test type curve values.

aquifer test modeling equations. Analyzing this effect requires the coupling of fluid flow and aquifer deformation (see Verruijt, 1969; Rodriques, 1983; Hsieh, 1996; Kim and Parizak, 1997; and Burbey, 1999).

STEHFEST ALGORITHM PUMPING TEST MATHEMATICAL MODELING EQUATIONS

Available Stehfest algorithm pumping test mathematical modeling equations cover infinite and finite aquifers, fully or partially penetrating wells with or without wellbore storage and skin, and the following aquifer conditions:

- Confined nonleaky
- Confined leaky without confining unit storativity
- Confined leaky with confining unit storativity and source unit above confining unit
- Confined leaky with confining unit storativity and impermeable unit above confining unit
- Confined leaky with confining unit storativity and water table within confining unit
- Confined fissure and block (double porosity) with slab-shaped blocks
- Confined fissure and block (double porosity) with sphere-shaped blocks
- Unconfined with or without delayed drainage at water table
- Induced streambed infiltration

Stehfest algorithm pumping test mathematical models are based on Laplace transforms $F(p)$ of groundwater functions $f(t)$ (Moench and Ogata, 1984, p. 150; Tien-Chang Lee, 1999, p. 80; Cheng, 2000, pp. 91–93). $f(t)$ can be calculated at any dimensional time $t > 0$ with a number of discrete $F(p)$ values using the following approximation equation:

$$f(t) \approx [(\ln 2)/t] \sum_{i=1}^{N} V_i F[i(\ln/2)/t)] \quad (3.99)$$

with

$$V_i = (-1)^{(N/2)+i} \sum_{k=(i+1)/2}^{\min(i,N/2)} [k^{N/2}(2k!)]/[(N/2 - k)!k!(k - 1)!(i - k)!(2k - i)!] \quad (3.100)$$

where ln2 = 0.693147180559945, t is the elapsed time, $F[i(\ln/2)/t)]$ is the appropriate groundwater flow Laplace transform equation in which $i(\ln/2)/t$ is substituted for the Laplace transform parameter p, N is the even number of Stehfest terms (4, 6, 8, etc.), ! is a factorial, and k is computed using integer arithmetic.

Type curve dimensionless drawdowns can be calculated at any dimensionless time by substituting r^2S/Tt or Tt/r^2S for t in Equation 3.99.

When too few or too many Stehfest terms are used, dimensionless drawdowns will be highly irregular (sometimes spikes will appear) especially in early time solutions. In this case, the number of Stehfest terms is changed and dimensionless drawdowns are recalculated. The sum of V_i values for any particular value of N is 0. Tables of V_i are presented by Walton (1996, pp. 62–64). Computer programs for calculating values of V_i are presented by Dougherty (1989, pp. 564–569), Moench (1994), Barlow and Moench (1999), and Cheng (2000). Since V_i depends only on N, it only needs to be calculated once for any value of N. The optimal value of N is 8 for pumping test and slugged well equations and 4 for derivative slugged well equations and slug observation well equations, assuming a personal computer and double precision calculations are used. Values of V_i for $N = 4$, $N = 6$, $N = 8$, and $N = 10$ are as follows:

$N = 4$

$V_i(1) = -0.2000000000000000D + 01$
$V_i(2) = 0.2600000000000000D + 02$
$V_i(3) = -0.4800000000000000D + 02$
$V_i(4) = 0.2400000000000000D + 02$

$N = 6$

$V_i(1) = 0.1000000000000000D + 01$
$V_i(2) = -0.4900000000000000D + 02$

$V_i(3) = 0.3660000000000000D + 03$
$V_i(4) = -0.8580000000000000D + 03$
$V_i(5) = 0.8100000000000000D + 03$
$V_i(6) = -0.2700000000000000D + 03$

$N = 8$
$V_i(1) = -0.3333333333333333D + 00$
$V_i(2) = 0.4833333333333333D + 02$
$V_i(3) = -0.9060000000000000D + 03$
$V_i(4) = 0.5464666666666667D + 04$
$V_i(5) = -0.1437666666666667D + 05$
$V_i(6) = 0.1873000000000000D + 05$
$V_i(7) = -0.1194666666666667D + 05$
$V_i(8) = 0.2986666666666667D + 04$

$N = 10$
$V_i(1) = 0.8333333333333333D - 01$
$V_i(2) = -0.3208333333333334D + 02$
$V_i(3) = 0.1279000000000000D + 04$
$V_i(4) = -0.1562366666666667D + 05$
$V_i(5) = 0.8424416666666666D + 05$
$V_i(6) = -0.2369575000000000D + 06$
$V_i(7) = 0.3759116666666667D + 06$
$V_i(8) = -0.3400716666666667D + 06$
$V_i(9) = 0.1640625000000000D + 06$
$V_i(10) = -0.3281250000000000D + 05$

Confined Aquifer Laplace Transform Equations

Laplace transform dimensionless drawdown equations for confined aquifers with specified aquifer boundary and well conditions are as follows (Moench and Ogata, 1984, pp. 153–168; Johns et al., 1992, p. 73; Cheng, 2000; Moench et al., 2001):

Infinite aquifer with fully penetrating wells without pumped and observation well delayed response (wellbore storage) and pumped wellbore skin (Theis equation):

$$\overline{h}_D = K_0(rk^{1/2})/p \tag{3.101}$$

where $rk^{1/2}$ is a factor (discussed below Equation 3.115) and p is the Laplace transform parameter.

Infinite aquifer with fully penetrating wells and pumped wellbore storage without observation well delayed response and pumped wellbore skin:

$$\overline{h}_D = K_0(rk^{1/2})/(p\ F_{wbs}) \tag{3.102}$$

$$F_{wbs} = W_D[(rk^{1/2})^2/r^2_D]K_0(rk^{1/2}/r_D) + (rk^{1/2}/r_D)K_1(rk^{1/2}/r_D) \tag{3.103}$$

where

$$W_D = (r_c^2 - r_p^2)/(2r_w^2 S) \quad (3.104)$$

$$r_D = r/r_w \quad (3.105)$$

r_c is the pumped well casing radius, r_p is the pump pipe radius, r_w is the pumped well effective radius, and S is the aquifer storativity.

Values of $K_1(x)$ are calculated with the following polynomial approximations presented by Abramowitz and Stegun (1964):

when $0 < x \leq 2$

$$K_1(x) = a_{10} + a_{11}[a_1 + (x^2/4)a_{12}] \quad (3.106)$$

$a_1 = 1.00000000$ $a_2 = 0.15443144$
$a_3 = -0.67278579$ $a_4 = 0.18156897$
$a_5 = -0.01919402$ $a_6 = 0.00110404$
$a_7 = -0.00004686$ $a_8 = a_6 + (x^2/4)a_7$
$a_9 = a_4 + (x^2/4)[a_5 + (x^2/4)a_8]$
$a_{10} = \log(x/2)I_1(x)$
$a_{11} = 1/x$
$a_{12} = \{a_2 + (x^2/4)[a_3 + (x^2/4)a_9]\}$

with $\text{abs}(x) < 3.75$

$$I_1(x) = x\{a_1 + (x/3.75)^2[a_2 + (x/3.75)^2 a_{10}]\} \quad (3.107)$$

$a_1 = 0.5$ $a_2 = 0.87890594$
$a_3 = 0.51498869$ $a_4 = 0.15084934$
$a_5 = 0.02658733$ $a_6 = 0.00301532$
$a_7 = 0.00032411$
$a_8 = a_6 + (x/3.75)^2 a_7$
$a_9 = a_4 + (x/3.75)^2[a_5 + (x/3.75)^2 a_8]$
$a_{10} = a_3 + (x/3.75)^2 a_9$

with $\text{abs}(x) \geq 3.75$

$$I_1(x) = [\exp(b_1)/(b_1)^{1/2}]a_{13} \quad (3.107a)$$

$a_1 = 0.39894228$ $a_2 = 0.03988024$
$a_3 = -0.00362018$ $a_4 = 0.00163801$
$a_5 = -0.01031555$ $a_6 = 0.02282967$

$$a_7 = -0.02895312 \qquad a_8 = 0.01787654$$
$$a_9 = -0.00420059$$
$$a_{10} = a_6 + b_2[a_7 + b_2(a_8 + b_2 a_9)]$$
$$a_{11} = a_5 + (b_2 a_{10})$$
$$a_{12} = a_4 + (b_2 a_{11})$$
$$a_{13} = a_1 + b_2[a_2 + b_2(a_3 + b_2 a_{12})]$$
$$b_1 = \text{abs}(x)$$
$$b_2 = 3.75/b_1$$

when $x > 2$

$$K_1(x) = [a_1 + (2/x)a_{10}][\exp(-x)/x^{0.5}] \tag{3.108}$$

$$a_1 = 1.25331414 \qquad a_2 = 0.23498619$$
$$a_3 = -0.0365562 \qquad a_4 = 0.01504268$$
$$a_5 = -0.00780353 \qquad a_6 = 0.00325614$$
$$a_7 = -0.00068245$$
$$a_8 = a_6 + (2/x)a_7$$
$$a_9 = a_4 + (2/x)[a_5 + (2/x)a_8]$$
$$a_{10} = a_2 + (2/x)[a_3 + (2/x)a_7]$$

Infinite aquifer with partially penetrating wells and pumped wellbore storage without observation well delayed response and pumped wellbore skin:

$$\overline{h}_D = [K_0(rk^{1/2}) + F_{pp}]/[p\ (F_{wbs})] \tag{3.109}$$

$$F_{pp} = 2/[(x_L - x_D)(x'_L - x'_D)] \sum_{n}^{\infty} 1/n^2\ [\sin(x_L n) - \sin(x_D n)]$$

$$[\sin(x'_L n) - \sin(x'_D n)][K_0[(rk^{1/2})^2 + (K_v/K_h)(n\pi r/b)^2]^{0.5} \tag{3.110}$$

where

$$x_L = \pi L/b \tag{3.111}$$

$$x_D = \pi D/b \tag{3.112}$$

$$x'_L = \pi L'/b \tag{3.113}$$

$$x'_D = \pi D'/b \tag{3.114}$$

K_v is the aquifer vertical hydraulic conductivity, K_h is the aquifer horizontal hydraulic conductivity, b is the aquifer thickness, L is the depth from the aquifer top to the pumped well base, D is the depth from the aquifer top to the top of the pumped well screen, L' is the depth from the aquifer top to the observation well base, D' is the depth from the aquifer top to the top of the observation well screen.

Finite aquifer with partially penetrating wells and pumped wellbore storage without observation well delayed response and pumped wellbore skin:

$$\overline{h}_D = [K_0(rk^{1/2}) + F_{pp} + K_0(r_{iw}k^{1/2})]/(p\ F_{wbs}) \tag{3.115}$$

The $rk^{1/2}$ factor is the Laplace–domain transform solution for a particular set of aquifer and real well conditions. The $r_{iw}k^{1/2}$ factor is the Laplace–domain transform solution for a particular set of aquifer and image well conditions. $K_0(\ldots)$ is the modified Bessel function of second kind and order zero. The Bessel function associated with a boundary image well is subtracted when there is a recharge boundary or is added when there is a barrier boundary. Additional Bessel functions associated with boundary image wells are added when there are several image wells. The effect of image well partially penetration is assumed to be negligible because the distance between the observation well and image well is usually large. The effects of boundary wellbore storage are assumed to be appreciable.

Finite aquifer with partially penetrating wells and pumped wellbore storage without observation well delayed response and with pumped wellbore skin:

$$\overline{h}_D = [K_0(rk^{1/2}) + (rk^{1/2})S_{wsf}K_1(rk^{1/2}) + F_{pp} + K_0(r_{iw}k^{1/2})]/(p\ F_{wbss}) \tag{3.116}$$

where

$$S_{wsf} = K_h d_s/(K_v\, r_w) \tag{3.117}$$

S_{wsf} is the wellbore skin factor, K_h is the aquifer horizontal hydraulic conductivity, K_v is the aquifer vertical hydraulic conductivity, r_w is the pumped well effective radius, and d_s is the skin thickness.

Finite aquifer with partially penetrating wells, pumped wellbore storage, observation well delayed response, and pumped wellbore skin (Moench et al., 2001, p. 12):

$$\overline{h}_{mD} = \overline{h}_D/(1 + W_{dp}p) \tag{3.118}$$

where $\overline{h}_{mD}$ is the Laplace transform dimensionless drawdown with observation well delayed response and W_{dp} is a dimensionless parameter defined as

$$W_{dp} = \pi r_o^2/(2\pi r_w^2 S_s F') \tag{3.119}$$

r_o is the observation well casing radius, r_w is the pumped well effective radius, $S_s = S/b$, and

$$F' = L_s/[\ln(x + 1 + x^2)^{0.5}] \tag{3.120}$$

$$x = (K_h / K_v)^{0.5} L_s / (2r_o) \tag{3.121}$$

K_h is the aquifer horizontal hydraulic conductivity, K_v is the aquifer vertical hydraulic conductivity, F' is a shape factor defined by Hvorslev (1951, case 8), L_s is the observation well or piezometer screen length, and r_o is the observation well radius. Delayed response in an observation piezometer cannot be simulated with Equation 3.119 because the screened length of the observation piezometer is 0 (Barlow and Moench, 1999, p. 12). Drawdown in an observation piezometer without delayed response is calculated based on the depth to the piezometer center. Drawdown in an observation piezometer with delayed response is simulated by assuming the observation piezometer is a partially penetrating observation well with a very short (1 ft) screen.

The $rk^{1/2}$ factors for commonly encountered confined aquifer conditions are:

Confined nonleaky aquifer (Moench and Ogata, 1984, p. 153):

$$rk^{1/2} = (r^2 Sp/T)^{0.5} \tag{3.122}$$

where r is the distance from the pumped well to the observation well, S is the aquifer storativity, p is the Laplace–domain variable, and T is the aquifer transmissivity.

Confined leaky aquifer without confining unit storativity and a source unit above the confining unit (Hantush, 1964, pp. 331–332):

$$rk^{1/2} = [r^2 Sp/T + (r/B)^2]^{0.5} \tag{3.123}$$

where

$$B = [T/(K'/b')]^{0.5} \tag{3.124}$$

r is the distance from the pumped well to the observation well, S is the aquifer storativity, p is the Laplace–domain variable, T is the aquifer transmissivity, K' is the confining unit vertical hydraulic conductivity, and b' is the confining unit thickness.

Confined leaky aquifer with confining unit storativity and the confining unit overlain by a source unit (Moench and Ogata, 1984, pp. 153–154):

$$rk^{1/2} = \{r^2 Sp/T + 4(r^2 Sp/T)^{0.5} \beta \coth[4(r^2 Sp/T)^{0.5} \beta/(r/B)^2]\}^{0.5} \tag{3.125}$$

where

$$\beta = [K'r/(4K_h b)][TS'/(SK'b')]^{0.5} \tag{3.126}$$

$$r/B = (r/b)[K'b/(K_h b')]^{0.5} \tag{3.127}$$

r is the distance from the pumped well to the observation well, S is the aquifer storativity, p is the Laplace–domain variable, T is the aquifer transmissivity, K_h is the aquifer horizontal hydraulic conductivity, b is the aquifer thickness, K' is the confining unit vertical hydraulic conductivity, b' is the confining unit thickness, and S' is the confining unit storativity.

Confined leaky aquifer with confining unit storativity and the confining unit overlain by an impermeable unit (Hantush, 1964, pp. 331–332):

$$rk^{1/2} = \{r^2Sp/T + 4(r^2Sp/T)^{0.5}\beta\tanh[4(r^2Sp/T)^{0.5}\beta/(r/B)^2]\}^{0.5} \quad (3.128)$$

where

$$\beta = [K'r/(4K_hb)][TS'/(SK'b')]^{0.5} \quad (3.129)$$

$$r/B = (r/b)[K'b/(K_hb')]^{0.5} \quad (3.130)$$

r is the distance from the pumped well to the observation well, S is the aquifer storativity, p is the Laplace–domain variable, T is the aquifer transmissivity, K' is the confining unit vertical hydraulic conductivity, K_h is the aquifer horizontal hydraulic conductivity, b is the aquifer thickness, b' is the confining unit thickness, and S' is the confining unit storativity.

Confined leaky aquifer with confining unit storativity and the confining unit contains the water table (Cooley and Case, 1973):

$$\begin{aligned} rk^{1/2} = \{&r^2Sp/T + 4(r^2Sp/T)^{0.5}\beta\tanh[4(r^2Sp/T)^{0.5}\beta/(r/b)^2] \\ &+ (r^2Sp/T)\operatorname{sech}^2[4(r^2Sp/T)^{0.5}\beta/(r/B)^2]/[(r^2Sp/T)(L_c/b)/(r/B)^2 \\ &+ (r/B)^2S/(16\beta^2S_{cY}) + (r^2Sp/T)/(4\beta)\tanh\,(4r^2Sp/T)\beta/(r/B)^2]\}^{0.5} \end{aligned} \quad (3.131)$$

where

$$\beta = [K'r/(4K_hb)][TS'/(SK'b')]^{0.5} \quad (3.132)$$

$$r/B = (r/b)\{K'b/[K_h(b' + L_{cf})]\}^{0.5} \quad (3.133)$$

r is the distance from the pumped well to the observation well, S is the aquifer storativity, p is the Laplace–domain variable, T is the aquifer transmissivity, b is the aquifer thickness, K' is the confining unit vertical hydraulic conductivity, K_h is the aquifer horizontal hydraulic conductivity, b' is the confining unit thickness, S' is the confining unit storativity, S_{cY} is the confining unit specific yield, and L_{cf} is the capillary fringe thickness.

Laplace transform dimensionless equations for an infinite confined leaky aquifer with fully penetrating wells without wellbore storage and skin, confining unit storativity, and a confining unit overlain by a variable head aquifer (two-aquifer system with drawdown in the unpumped aquifer) are as follows (Neuman and Witherspoon, 1969a, 1969b; see Moench and Ogata, 1984):

For the pumped aquifer

$$\overline{h}_D = 1/p\ [(A_2 - k_1)/D_{vh}]\ K_0(rk_1^{1/2}) - 1/p[(A_2 - k_2)/D_{vh}]\ K_0(rk_2^{1/2}) \quad (3.134)$$

where

$$k_1 = 1/2[A_1 + A_2 - D_{vh}] \quad (3.135)$$

$$k_2 = 1/2[A_1 + A_2 + D_{vh}] \quad (3.136)$$

$$D_{vh} = [4B_1B_2 + (A_1 - A_2)^2]^{0.5} \quad (3.137)$$

$$A_1 = r^{-2}[\eta^2 + 4\eta\beta_{11}\coth(\psi)] \quad (3.138)$$

$$A_2 = r^{-2}[\eta^2(\alpha_1/\alpha_2) + 4\eta\beta_{21}\ (\alpha_1/\alpha_2)^{0.5}\coth(\psi)] \quad (3.139)$$

$$B_1 = r^{-2}\ 4\eta\beta_{11}\ [1/\sinh(\psi)] \quad (3.140)$$

$$B_2 = r^{-2}\ 4\eta\beta_{21}\ (\alpha_1/\alpha_2)^{0.5}\ [1/\sinh(\psi)] \quad (3.141)$$

$$\psi = 4\eta\beta_{11}\ (r/B_{11})^{-2} \quad (3.142)$$

$$\beta_{11} = 1/4\ K'/K_1\ r/b_1\ (\alpha_1/\alpha')^{0.5} \quad (3.143)$$

$$r/B_{11} = r/b_1\ (K'/K_1\ b_1/b')^{0.5} \quad (3.144)$$

$$\beta_{21} = \beta_{11}\ T_1/T_2\ (\alpha_2/\alpha_1)^{0.5} \quad (3.145)$$

$$r/B_{21} = r/B_{11}\ (T_1/T_2)^{0.5} \quad (3.146)$$

$$\alpha_1 = T_1/S_1 \quad (3.147)$$

$$\alpha_2 = T_2/S_2 \quad (3.148)$$

$$\alpha' = K'b'/S' \quad (3.149)$$

$$\eta = [(r^2S_1/T_1)\ p]^{0.5} \quad (3.150)$$

For the unpumped aquifer

$$\overline{h}_{D2} = [B_2/(pD_{vh})]\ [K_0(rk_1^{1/2}) - K_0(rk_2^{1/2})] \quad (3.151)$$

For the confining unit

$$\bar{h}'_D = [\sinh(\psi z/b')/\sinh(\psi)]\, \bar{h}_{D2} + \{\sinh[\psi(1 - z/b')]/\sinh(\psi)\}\bar{h}_{D1} \quad (3.152)$$

For two aquifers with identical hydraulic properties

$$\bar{h}_{D1} = 1/(2p)\ [K_0(rk_1^{1/2}) + K_0(rk_2^{1/2})] \quad (3.153)$$

where p is the Laplace–domain variable, r is the distance from the pumped well to the observation well, b_1 is the pumped aquifer thickness, b' is the confining unit thickness, K_1 is the pumped aquifer horizontal hydraulic conductivity, K' is the confining unit vertical hydraulic conductivity, r is the distance from the pumped well to the observation well, T_1 is the pumped aquifer transmissivity, T_2 is the unpumped aquifer transmissivity, S_1 is the pumped aquifer storativity, S_2 is the unpumped aquifer storativity or specific yield, and S' is the confining unit storativity.

The effects of drawdown in the unpumped aquifer may not be appreciable during the short duration of most aquifer tests, but, these effects can be quite significant over longer periods of time.

Dennis and Motz (1998) extended the Neuman and Witherspoon (1969b) two-aquifer system equations to cover pumping as well as reduction in evapotranspiration from the aquifer above the confining unit. These equations are included in their Fortran program NSSCON. Cheng (2000) extended the Neuman and Witherspoon (1969b) two-aquifer system equations to cover multiple-aquifer-confining unit systems. These equations are included in a suite of Fortran programs. Both the Dennis and Motz and Cheng equations are for infinite aquifers and fully penetrating wells with no wellbore storage and skin.

Moench (1985) extended the Hantush (1960) theory of a confined leaky aquifer overlain and underlain by confining units to cover wellbore storage and skin. The upper boundary of the overlying confining unit and the lower boundary of the underlying confining unit can be constant head or no-flow boundaries.

Confined fissure and block aquifer (double porosity) wells are fully penetrating. The fissure has a skin; the block is leaky with storativity and is overlain by a source unit (Moench, 1984, pp. 831–846).

For a slab-shaped block:

$$rk^{1/2} = \{(r^2Sp/T) + r_{Df}^{\,2}(m_f)[\tanh(m_f)]/[1 + S_f(m_f)\tanh(m_f)]\}^{0.5} \quad (3.154)$$

For a sphere-shaped block:

$$rk^{1/2} = \{(r^2Sp/T) + 3r_{Df}^{\,2}[(m_f)\coth(m_f) - 1]/\{1 + S_f[m_f\coth(m_f) - 1]\}^{0.5} \quad (3.155)$$

where

$$r_{Df} = [r/(b'_b/2)][K'_b/K_f]^{0.5} \quad (3.156)$$

$$m_f = (S_r p)^{0.5}/r_{Df} \quad (3.157)$$

$$S_r = S'_b/S_s \quad (3.158)$$

$$S_f = K'_b b_s/[K_f(b'_b/2)] \quad (3.159)$$

T is the fissure transmissivity, K_f is the fissure horizontal hydraulic conductivity, K'_f is the block vertical hydraulic conductivity, b'_b is the average block thickness between fissure zones, S'_b is the block specific storage, S_f is the fissure specific storage, S is the aquifer storativity, p is the Laplace–domain variable, K_s is the fissure skin hydraulic conductivity, r is the distance from the pumped well to the observation well, b_s is the fissure skin thickness. The aquifer is assumed to consist of two interacting, overlapping continua: a continuum of low-hydraulic conductivity, primary porosity blocks and a continuum of high-hydraulic conductivity, secondary porosity fissures. The parameters of the continuums are homogeneous and isotropic. The double-porosity aquifer is confined above and below by impermeable formations. Groundwater enters a single pumped wellbore through the fissures and not the block. The discharge rate is constant. There is no initial groundwater flow. There is transient flow from blocks to fissures causing type curves to show a transition from early to late time. The length of the transition time is controlled by S_r and the vertical position of the transient period is controlled by r_D. Slab-shaped blocks are usually assumed. Closely spaced water entries are needed to justify the use of sphere-shaped blocks. S_f is typically 1/10 to 1/100 of S'_b. K_f is typically 0.001 to 10 ft/day and K'_b is typically 1E-7 to 1E-4 ft/day. T, S'_b, and S_f are associated with the combined fissure and block thickness.

Note that the $rk^{1/2}$ factors for leaky and fissure and block aquifers are equal to the $rk^{1/2}$ factor for the nonleaky aquifer plus source terms.

Unconfined Aquifer Laplace Transform Equations

Laplace transform dimensionless drawdown equations for unconfined aquifers with specified aquifer and well conditions are as follows (Moench, 1997, 1998; Moench et al., 2001 and Addendum):

For a pumped well:

$$\overline{h}_D = 2(A + S_{wsf})/\{p(1_D - d_D)[1 + W_d p(A + S_{wsf})]\} \quad (3.160)$$

where

$$A = 2/(l_D - d_D)\sum_{n=0}^{\infty} K_0(q_n)\{\sin[n\pi(1 - d_D)] - \sin[n\pi(1 - l_D)]\}^2/$$
$$\{n\pi q_n K_1(q_n)\} \quad (3.161)$$

$$l_D = L/b \quad (3.162)$$

$$d_D = D/b \tag{3.163}$$

$$W_d = \pi r_{ce}^2/[2\pi r_w^2 S_s(L - D)] \tag{3.164}$$

$$q_n = (\varepsilon_n^2 \beta_w + p)^{0.5} \tag{3.165}$$

ε_n, where $n = 0, 1, 2, \ldots$ are the roots of

$$\varepsilon_n \tan(\varepsilon_n) = p/M \sum_{n=1}^{M} [1/(\sigma\beta_w + p/\gamma_m)] \tag{3.166}$$

M is the number of empirical constants for gradual drainage from the unsaturated zone, $\sigma = S/S_y$, and S is the aquifer storativity and S_y is the aquifer specific yield:

$$\gamma_m = \alpha_m b S_y/K_v \tag{3.167}$$

α_m is the mth empirical constant for gradual drainage from the unsaturated zone.

$$\beta_w = K_D r_{wD}^2 \tag{3.168}$$

$$K_D = K_v/K_h \tag{3.169}$$

$$r_{wD} = r_w/b \tag{3.170}$$

$$\beta = \beta_w r_D^2 \tag{3.171}$$

$$r_D = r/r_w \tag{3.172}$$

$$r_{ce} = (r_c^2 - r_p^2)^{0.5} \tag{3.173}$$

S_{wsf} is the wellbore skin factor = $K_h d_s/(K_v r_w)$, K_h is the aquifer horizontal hydraulic conductivity, d_s is the skin thickness (for simplicity, drawdown due to skin is presumed to increase linearly with the discharge rate, Tien-Chang Lee, 1999, p. 181), K_v is the aquifer vertical hydraulic conductivity, r_w is the pumped well effective radius, S_s is the aquifer specific storativity, b is the aquifer thickness, r is the distance from the pumped well to the observation well, r_c is the pumped well casing radius, r_p is the pump pipe radius, p is the Laplace–domain variable, L is the depth from the aquifer top to the pumped well base, and D is the depth from the aquifer top to the top of the pumped well screen.

For an observation well:

$$\bar{h}_D = 2E/\{p(l_D - d_D)[1 + W_d p(A + S_{wsf})]\} \tag{3.174}$$

where

$$E = 2\sum_{n=0}^{\infty} K_0(q_n r_D)\{\sin[n\pi(1 - d_D)] - \sin[n\pi(1 - l_D)]\}/$$
$$\{n\pi q_n K_1(q_n)[n\pi + 0.5\sin(2n\pi)]\}$$
$$[\sin(n\pi z_{D2}) - \sin(n\pi z_{D1})]/(z_{D2} - z_{D1}) \quad (3.175)$$

$$l_D = L/b \quad (3.176)$$

$$d_D = D/b \quad (3.177)$$

$$z_{D1} = z_1/b \quad (3.178)$$

$$z_{D2} = z_2/b \quad (3.179)$$

$$W_d = \pi r_{ce}^2/[2\pi r_w^2 S_s(L - D)] \quad (3.180)$$

$$q_n = (\varepsilon_n^2 \beta_w + p)^{0.5} \quad (3.181)$$

ε_n, where $n = 0, 1, 2, \ldots$ are the roots of

$$\varepsilon_n \tan(\varepsilon_n) = p/M \sum_{n=1}^{M} [1/(\sigma\beta_w + p/\gamma_m)] \quad (3.182)$$

M is the number of empirical constants for gradual drainage from the unsaturated zone, $\sigma = S/S_y$, and S is the aquifer storativity and S_y is the aquifer specific yield:

$$\gamma_m = \alpha_m b S_y / K_v \quad (3.183)$$

α_m is the mth empirical constant for gradual drainage from the unsaturated zone.

$$\beta_w = K_D r_{wD}^2 \quad (3.184)$$

$$K_D = K_v / K_h \quad (3.185)$$

$$r_{wD} = r_w / b \quad (3.186)$$

$$q_n r_D = (\varepsilon_n^2 \beta + p r_D^2)^{0.5} \quad (3.187)$$

$$\beta = \beta_w r_D^2 \quad (3.188)$$

$$r_D = r/r_w \quad (3.189)$$

$$r_{ce} = (r_c^2 - r_p^2)^{0.5} \quad (3.190)$$

S_{wsf} is the wellbore skin factor = $K_h d_s/(K_v r_w)$, K_h is the aquifer horizontal hydraulic conductivity, d_s is the skin thickness (for simplicity, drawdown due to skin is presumed to increase linearly with the discharge rate (Tien-Chang Lee, 1999, p. 181), K_v is the aquifer vertical hydraulic conductivity, r_w is the pumped well effective radius, S_s is the aquifer specific storativity, b is the aquifer thickness, r is the distance from the pumped well to the observation well, r_c is the pumped well casing radius, r_p is the pump pipe radius, p is the Laplace–domain variable, 1 is the depth from the aquifer top to the pumped well base, d is the depth from the aquifer top to the top of the pumped well screen, z_2 is the depth from the aquifer top to the observation well base, and z_1 is the depth from the aquifer top to the top of the observation well screen.

For a piezometer:

$$\overline{h}_D = 2E/\{p(l_D - d_D)[1 + W_d p(A + S_{wsf})]\} \tag{3.191}$$

where

$$E = 2\sum_{n=0}^{\infty} K_0(q_n r_D)\{\sin[n\pi(1 - d_D)] - \sin[n\pi(1 - l_D)]\}/$$
$$[n\pi q_n K_1(q_n)] \cos(n\pi z_D) \tag{3.192}$$

$$l_D = Ll/b \tag{3.193}$$

$$d_D = D/b \tag{3.194}$$

$$z_D = z_p/b \tag{3.195}$$

$$q_n = (\varepsilon^2{}_n\beta_w + p)^{0.5} \tag{3.196}$$

$$q_n r_D = (\varepsilon^2{}_n\beta + pr_D{}^2)^{0.5} \tag{3.197}$$

ε_n, where $n = 0, 1, 2, \ldots$ are the roots of

$$\varepsilon_n \tan(\varepsilon_n) = p/M\sum_{n=1}^{M} [1/(\sigma\beta_w + p/\gamma_m)] \tag{3.198}$$

M is the number of empirical constants for gradual drainage from the unsaturated zone, $\sigma = S/S_y$, and S is the storativity and S_y is the specific yield:

$$\gamma_m = \alpha_m bS_y/K_v \tag{3.199}$$

α_m is the mth empirical constant for gradual drainage from the unsaturated zone.

$$W_d = \pi r_{ce}^2/[2\pi r_w^2 S_s(L - D)] \tag{3.200}$$

$$\beta_w = K_D r_{wD}^2 \tag{3.201}$$

$$K_D = K_v / K_h \tag{3.202}$$

$$r_{wD} = r_w/b \tag{3.203}$$

$$\beta = \beta_w r_D^2 \tag{3.204}$$

$$r_D = r/r_w \tag{3.205}$$

$$r_{ce} = (r_c^2 - r_p^2)^{0.5} \tag{3.206}$$

S_{wsf} is the wellbore skin factor = $K_h d_s/(K_s r_w)$, K_h is the aquifer horizontal hydraulic conductivity, d_s is the skin thickness (for simplicity, drawdown due to skin is presumed to increase linearly with the discharge rate (Tien-Chang Lee, 1999, p. 181), K_v is the aquifer vertical hydraulic conductivity, r_w is the pumped well effective radius, S_s is the aquifer specific storativity, b is the aquifer thickness, r is the distance from the pumped well to the observation well, r_c is the pumped well casing radius, r_p is the pump pipe radius, p is the Laplace–domain variable, L is the depth from the aquifer top to the pumped well base, and D is the depth from the aquifer top to the top of the pumped well screen, and z_p is the vertical distance above the base of the aquifer to the center of the piezometer screen. The process of gradual drainage from the unsaturated zone above the water table, effects of well partial penetration, pumped wellbore storage, and pumped well skin are simulated. Well loss due to the turbulent flow near the pumped well is not simulated.

Induced Streambed Infiltration Fourier–Laplace Transform Equations

Induced streambed infiltration Fourier–Laplace transform dimensionless drawdown equations for unconfined and confined leaky aquifers are presented by Butler et al. (2001); Butler and Tsou (2001); and Zhan and Butler (2005). These equations assume negligible wellbore storage and skin, fully penetrating wells, and finite width partially penetrating streambeds. Equations for unconfined aquifers (Butler and Tsou, 2001) are as follow:

Beyond the streambed in Zone 1:

$$\overline{\Phi}_1 (\varepsilon, \omega, p) = (T_f)[e^a + e^b] \tag{3.207}$$

Beneath the streambed in Zone 2:

$$\bar{\bar{\Phi}}_2\ (\varepsilon, \omega, p) = (T_f)[(A_1)e^c + (B_1)e^d] \tag{3.208}$$

Between pumped well and streambed in Zone 3:

$$\bar{\bar{\Phi}}_3\ (\varepsilon, \omega, p) = (T_f)[(D_1)e^f + (E_1)e^g],\ 0 \leq \varepsilon \leq \alpha \tag{3.209}$$

Between pumped well and right boundary in Zone 3:

$$\bar{\bar{\Phi}}_3\ (\varepsilon, \omega, p) = [(T_f)(G_1)/(H_1)][e^f + e^h],\ \alpha < \varepsilon \leq X_{RB} \tag{3.210}$$

where

$$\Phi_i\ \text{(dimensionless drawdown)} = s_i T_3/Q,\ i = 1, 3 \tag{3.211}$$

$$\tau\ \text{(dimensionless time)} = (T_3 t)/(w^2 S_3) \tag{3.212}$$

$$\xi = x/w \tag{3.213}$$

$$\eta = y/w \tag{3.214}$$

$$\alpha = a/w \tag{3.215}$$

$$B\ \text{(stream leakance)} = (k' w^2)/(b' T_2) \tag{3.216}$$

$$X_{RB} = x_{rb}/w \tag{3.217}$$

$$X_{LB} = x_{lb}/w \tag{3.218}$$

$$\gamma_i = T_{i+1}/T_i\ i = 1, 2 \tag{3.219}$$

$$P_i = \mu_i/\mu_3\ i = 1, 2 \tag{3.220}$$

$$\mu_i = S_i/T_i\ i = 1, 3 \tag{3.221}$$

$$\lambda_1 = (\omega^2 + P_1 p)^{0.5} \tag{3.222}$$

$$\lambda_2 = (\omega^2 + B + P_2 p)^{0.5} \tag{3.223}$$

$$\lambda_3 = (\omega^2 + p)^{0.5} \tag{3.224}$$

$$A_1 = 1/2(e^r + e^{r1}) + [\lambda_1/(/(2\ \gamma_1\ \lambda_2)][e^r - e^{r1}] \tag{3.225}$$

$$B_1 = 1/2(e^{r2} + e^{r3}) - [\lambda_1/(/(2\ \gamma_1\ \lambda_2)]\ [e^{r2} - e^{r4}] \tag{3.226}$$

$$D_1 = 1/2[(A_1) + (B_1)] + [\lambda_1/(2\gamma_2\lambda_3)]\ [(A_1) - (B_1)] \tag{3.227}$$

$$a = 2\lambda_1 X_{\Lambda B} + \lambda_1\xi \tag{3.228}$$

$$b = -\lambda_1\xi \tag{3.229}$$

$$c = \lambda_2\xi \tag{3.230}$$

$$d = -\lambda_2\xi \tag{3.231}$$

$$f = \lambda_3\xi \tag{3.232}$$

$$g = -\lambda_3\xi \tag{3.233}$$

$$h = 2\ \lambda_3 X_{RB} - \lambda_3\xi \tag{3.234}$$

$$r = 2\ \lambda_1 X_{LB} - \lambda_1 + \lambda_2 \tag{3.235}$$

$$r1 = \lambda_1 + \lambda_2 \tag{3.236}$$

$$r2 = 2\ \lambda_1 X_{LB} - \lambda_1 - \lambda_2 \tag{3.237}$$

$$r3 = \lambda_1 - \lambda_2 \tag{3.238}$$

$$r4 = -\lambda_1 - \lambda_2 \tag{3.239}$$

$$E_1 = 1/2[(A_1) + (B_1)] - [\lambda_2/(2\ \gamma_2\ \gamma_3)][(A_1) - (B_1)] \tag{3.240}$$

$$F_1 = -1/[\ \lambda_3 p(2\pi)^{0.5}] \tag{3.241}$$

$$G_1 = (D_1)e^{r5} + (E_1)e^{r6} \tag{3.242}$$

$$r5 = \lambda_3\alpha \tag{3.243}$$

$$r6 = -\lambda_3\alpha \tag{3.244}$$

$$H_1 = e^{r5} + e^{r7} \tag{3.245}$$

$$r7 = 2\lambda_3 X_{RB} - \lambda_3\alpha \tag{3.246}$$

$$J_1 = (D_1)e^{r5} - (E_1)e^{-r5} \tag{3.247}$$

$$K_1 = e^{r5} - e^{r7} \tag{3.248}$$

$$T_f = (F_1)(H_1)/[(G_1)(K_1) - (J_1)(H_1)] \tag{3.249}$$

$\overline{\overline{\Phi}}_i$ = Fourier–Laplace transform of Φ_i, i = 1,3; p = Laplace–transform variable; ω = Fourier transform variable; s_i is the drawdown in Zone i; T_i is the aquifer transmissivity in Zone i; Q is the discharge rate; t is the elapsed time; w is the streambed width; S_i is the aquifer storativity in Zone i; x is the X coordinate; y is the Y coordinate; a is the distance from the streambed to the pumped well; k is the streambed vertical hydraulic conductivity; b is the streambed thickness; x_{rb} is the distance from the right boundary to the right side of the streambed; x_{lb} is the distance from the left boundary to the right side of the streambed; the streambed bank nearest the pumped well is the zero X-coordinate baseline; X coordinates are positive to the right of the baseline and negative to the left of the baseline; the line at a right angle to the streambed through the pumped well is the zero Y-coordinate baseline; Y coordinates are positive above the baseline and negative below the baseline; parallel barrier boundaries occur to the right and left of the streambed baseline; distances to the barrier boundaries are positive to the right of the streambed baseline and negative to the left of the streambed baseline; the effects of the barrier boundaries become negligible when the distances from the streambed baseline to the boundaries are large enough (10,000 ft); Zone 1 refers to the aquifer left beyond the streambed from the pumped well; Zone 2 refers to the aquifer beneath the streambed; and Zone 3 refers to the aquifer to the right of the streambed. The equations given above are most readily evaluated using a numerical scheme. A Mathematica® add-on package (Mallet, 2000) can be used for the joint Fourier–Laplace numerical inversion.

The Laplace solution for stream depletion is (Butler and Tsou, 2001):

$$\Delta\overline{Q}(p) = B/(\gamma_2\lambda^*_2)(T_f)[(A_1)(1 - e^{-r8}) - (B_1)(1 - e^{r8})] \tag{3.250}$$

where

$$r8 = (B + P_2p)^{0.5} \tag{3.251}$$

$$P_2 = (S_2/T_2)/(S_3/T_3) \tag{3.252}$$

Butler and Tsou (2000) developed the Fortran analytical program StrpStrm for calculating time-drawdown values with induced streambed infiltration in unconfined aquifers. The program can be downloaded at www.kgs.ku.edu/StreamAq/Software/strp.html.

Hunt (1999) presents Fourier–Laplace and other analytical equations for calculating time-drawdown and stream depletion values with induced streambed infiltration in unconfined aquifers. These equations assume negligible wellbore storage and skin, fully penetrating wells, and finite width partially penetrating streambeds.

Fox and Durnford (2002) developed the Fortran analytical program STRMAQ for calculating time-drawdown values with induced streambed infiltration in unconfined and confined nonleaky aquifers. STRMAQ uses analogous well functions to combine WTAQ (Barlow and Moench, 1999) and induced streambed infiltration equations presented by Hunt (1999). STRMAQ requires three files: filename, input, and output. STRMAQ accounts for well partial penetration, wellbore skin, wellbore storage, and finite width partially penetrating streambeds. STRMAQ can be downloaded at www.engr.colostate.edu/~durnford/projects/ NAPL/STRMAQ_Readme.txt.

Wellbore Skin Effects

Effects of any well skin (see Figure 3.4) should be considered in estimating the pumped well effective radius (Moench et al., 2002, p. 18). If the well skin is less permeable than the aquifer, drawdown in the pumped well is increased (the effective radius decreases) and there is an apparent increase in wellbore storage, which reduces drawdowns in the aquifer at early times. If the well skin is more permeable than the aquifer, drawdown in the pumped well is decreased (the effective radius increases) and there is an apparent decrease in wellbore storage, which increases drawdowns in the aquifer at early times.

The pumped well effective radius can be estimated by calculating pumped well drawdowns for a selected time based on aquifer parameter values estimated with observation well data and several trial effective radius values. Calculated drawdowns are compared with the measured drawdown for the selected time. The trial effective radius that results in a match between calculated and measured drawdowns is assigned to the pumped well.

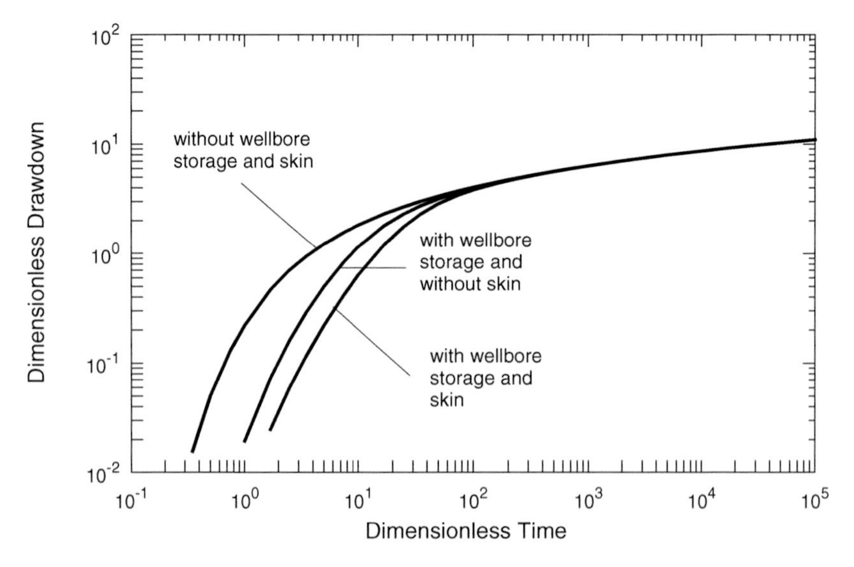

FIGURE 3.4 Graph showing wellbore storage and skin effects on pumping test type curve values.

STEHFEST ALGORITHM SLUG TEST MATHEMATICAL MODELING EQUATIONS

Several Stehfest algorithm slug test mathematical modeling equations are available (Dougherty, 1989; Novakowski, 1989; and Hyder et al., 1994). In addition, the general relationship between pumping test and slug test models (Ramey and Agarwal, 1972; Peres et al., 1989) can be used to generate Stehfest algorithm slug test mathematical modeling equations with Stehfest algorithm pumping test mathematical modeling equations for partially penetrating wells in confined nonleaky, confined leaky, confined fissure and block, and unconfined aquifer conditions. Under favorable conditions when the magnitude of normalized heads during late time portions of the slug test are appreciable, it is possible to estimate the confining unit vertical hydraulic conductivity under confined leaky aquifer conditions and the vertical to horizontal hydraulic conductivity ratio under unconfined aquifer conditions.

CONFINED AQUIFER LAPLACE–TRANSFORM EQUATIONS

The slug test Laplace–transform solution for a slugged well in an infinite confined nonleaky aquifer with partially penetrating wells and wellbore storage is as follows (Dougherty, 1989, pp. 567–568):

$$\overline{h}_D/H_o = C_D\rho(A_o + A)/[1 + pC_D\rho(A_o + A)] \quad (3.253)$$

where

$$A_o = K_0(p^{1/2})/[p^{1/2}K_1(p^{1/2})] \quad (3.254)$$

$$A = 2/\rho^2 \sum_{n=1}^{\infty} \{[\sin(n\pi z_{bD}/b_D) - \sin(n\pi z_{aD}/b_D)]^2\, K_0[(p + n^2\, \pi^2/b_D{}^2)^{0.5}]\}/ \{n^2\pi^2\, (p + n^2\pi^2/b_D{}^2)^{0.5}\, K_1[(p + n^2\, \pi^2/b_D{}^2)^{0.5}]\} \quad (3.255)$$

$$C_D = r_c{}^2/(2\, r_w{}^2\, S\, \rho) \quad (3.256)$$

$$\rho = (z_b - z_a)/b \quad (3.257)$$

$$z_{aD} = z_a/r_w \quad (3.258)$$

$$z_{bD} = z_b/r_w \quad (3.259)$$

$$b_D = b/r_w \quad (3.260)$$

r_w is the pumped well effective radius, S is the aquifer storativity, r_c is the pumped well casing radius, p is the Laplace–domain variable, H_o is the initial displacement from static head, z_a is the distance from the aquifer base to the slugged well screen base, z_b is the distance from the aquifer base to the slugged well screen top, and b is the aquifer thickness.

The slug test Laplace–transform solution for an observation well in an infinite confined nonleaky aquifer with fully penetrating wells and no observation well-bore storage is as follows (Novakowski, 1989, p. 2379):

$$\overline{h}_D/H_o = K_0(r_D p^{1/2})/\{p^{1/2}[p^{1/2}K_0(p^{1/2}) + (1/C_D)K_1(p^{1/2})]\} \tag{3.261}$$

where

$$r_D = r^2/r_w^2 \tag{3.262}$$

r is the distance between the slugged and observation wells and r_w is the pumped well effective radius.

A numerical inversion program TYPCURV was developed by Novakowski (1990) to generate slug test observation well dimensionless time-normalized head values.

Pumping-Slug Test Relationship

Pumping and slug test responses are related by the following dimensionless equation (Ramey and Agarwal, 1972; Peres et al., 1989):

$$H/H_0(t_D,r_D,C_D) = C_D[dp_D/dt_D(t_D,r_D,C_D)] \tag{3.263}$$

where

$$t_D = Tt/(r^2S) \tag{3.264}$$

$$r_D = r/r_w \tag{3.265}$$

$$C_D = r_c^2/(2r_w^2S) \tag{3.266}$$

$$p_D = 2\pi Ts/Q \tag{3.267}$$

T is the aquifer transmissivity, t is the elapsed time, S is the aquifer storativity, p_D is the pumping test dimensionless drawdown, H/H_0 is the slug test normalized head, r_w is the slugged well effective radius, r_c is the slugged well casing radius, r is the distance between the axis of the pumped or slugged well and the observation point (r_w is substituted for r in the case of the pumped or slugged well), $dp_D/dt_D(t_D,r_D,C_D)$ is the first derivative of the dimensionless pumping test drawdown with respect to the first derivative of the dimensionless time t_D, s is the drawdown, and Q is the pumped well constant discharge rate.

Thus, slug test time-normalized heads are first derivatives (slopes) of the dimensionless pumping test time drawdowns calculated with Stehfest algorithm pumping test mathematical modeling equations multiplied by C_D. First derivatives of pumping test dimensionless time drawdown are calculated with an algorithm listed by Bourdet et al. (1989). That algorithm calculates the first derivative of head change with respect to the natural logarithm of the change in time. The derivative is averaged over time periods before and after the point of interest. The slopes of the head change vs. the change in time are weighted and the head derivative for the point of interest is calculated with the following equation:

$$(dH/dt)_i = [(\Delta H_1/\Delta t_I)\Delta t_2 + (\Delta H_2/\Delta t_2)\Delta t_1]/(\Delta t_1 + \Delta t_2) \tag{3.268}$$

where subscript 1 refers to the points before the points of interest i, subscript 2 refers to the points after the points of interest i, $(dH/dt)_i$ are the slopes of the pumped or observation well dimensionless head changes vs. the changes in time at the points of interest, ΔH_1 is the pumped or observation well dimensionless head change over the interval between the point of interest i and the point before the point of interest, ΔH_2 is the pumped or observation well dimensionless head change over the interval between the point of interest i and the point after the point of interest, Δt_1 is the dimensionless natural logarithmic time change over the interval between the point of interest i and the point before the point of interest, and Δt_2 is the dimensionless natural logarithmic time change over the interval between the point of interest i and the point after the point of interest.

Two derivative algorithm methods are supported in the program DERIV developed by Spane and Wurster (1993): fixed endpoint and least-squares fit. The fixed-endpoint method is usually used for calculating derivatives of type curve values that are relatively free of noise. The least-squares fit method is usually used for calculating derivatives of noisy test data. In the fixed-endpoint method, the points immediately before and after the specified time L spacing from the point of interest are used in calculating mean slopes. The calculated slopes from the fixed endpoints to the point of interest are then weighted by multiplying each by its time distance to the point of interest, divided by the sum of the time distances to the two endpoints.

In the least-squares method, all data from the points immediately before and after the specified L spacing are used in calculating the slopes to the left and right of the point of interest. The calculated slopes are then weighted as described for the fixed-endpoint method. The L spacing may range from 0 to 5 (Spane and Wurster, 1993). An L-spacing value of 0.2 is commonly used to reduce noise in the calculated derivative. Larger L-spacing values can lead to oversmoothing of data.

Aquifer and Well Condition Effects

Aquifer type, well penetration, wellbore skin, and observation delayed response (wellbore storage) can appreciably affect slug test time-normalized heads. Slugged well time-normalized heads for different aquifer types (confined nonleaky, confined

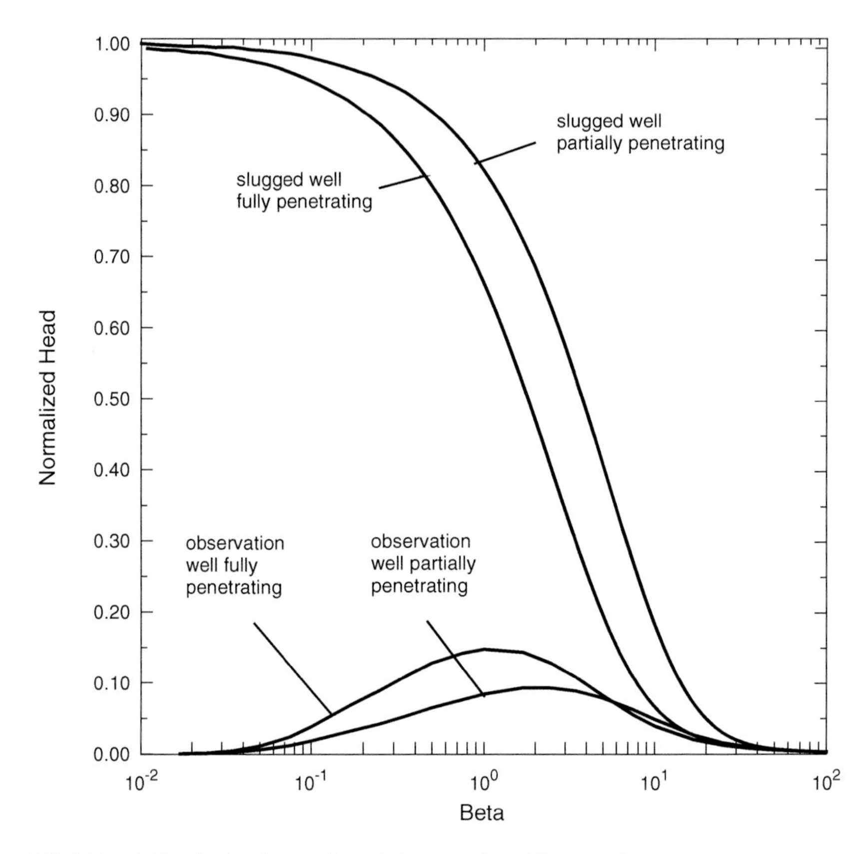

FIGURE 3.5 Graph showing well partial penetration effects on slug test type curve values.

leaky, and confined fissure and block aquifers) with the same parameter values are nearly identical except for late times. Slugged well time-normalized heads for unconfined aquifers are appreciably offset from time-normalized heads for other aquifer types. Slugged well time-normalized heads for various ratios of aquifer vertical to horizontal hydraulic conductivity are similar in shape and closely spaced.

As demonstrated in Figure 3.5, time-normalized heads for a partially penetrating slugged well or a nearby observation well are shifted to the right of time-normalized heads for a fully penetrating slugged well. It is apparent that erroneously low hydraulic conductivity values are calculated by applying fully penetrating slugged well models to data for partially penetrating slugged wells.

Slugged well time-normalized heads can be significantly affected by wellbore skin. The hydraulic conductivity of the wellbore skin can either be larger or smaller than that of the formation. As demonstrated in Figure 3.6, time-normalized heads for a slugged well or a nearby observation well with wellbore skin whose hydraulic conductivity is less than that of the formation are shifted to the right of time-normalized heads for a slugged well with no wellbore skin. Erroneously

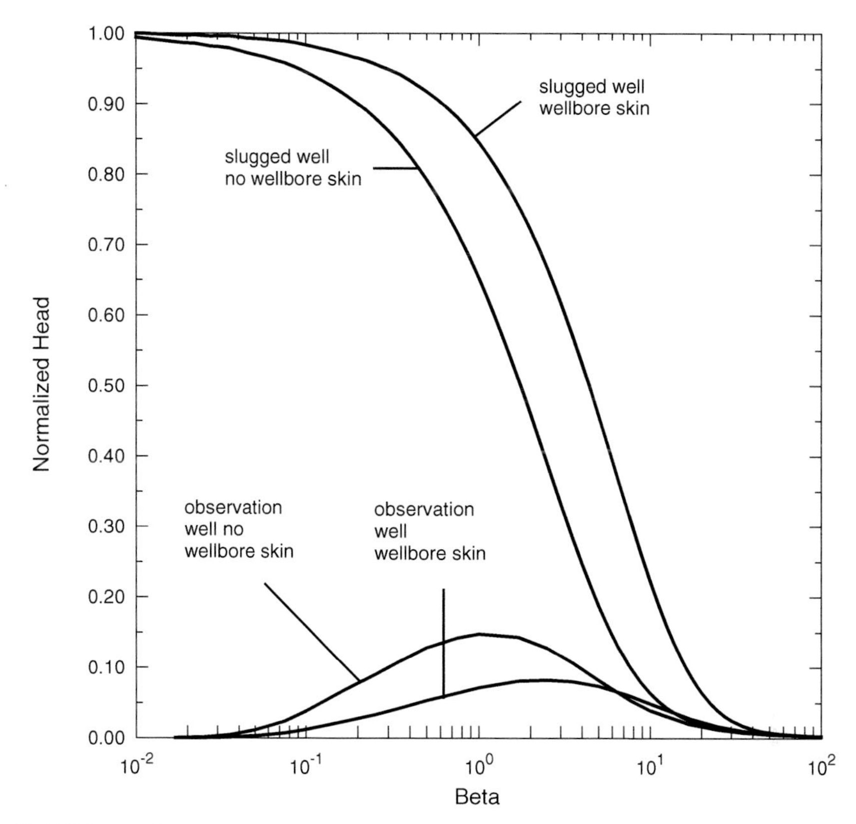

FIGURE 3.6 Graph showing wellbore skin effects on slug test type curve values.

low hydraulic conductivity values are calculated by applying slugged well with no wellbore skin models to data for slugged wells with wellbore skin. An implausible low storativity estimate obtained with a Stehfest algorithm slug test model indicates the presence of wellbore skin.

Moench and Hsieh (1985, p. 20) present an equation for analysis of slugged well test data accounting for a skin of finite thickness. The equation assumes a confined nonleaky aquifer and a fully penetrating well. Families of type curves generated with that equation for different values of the ratio of the aquifer hydraulic conductivity to the skin hydraulic conductivity have nearly identical shapes except for very low ratios. Therefore, there will be a large degree of nonuniqueness in matching test data to a family of type curves. Accurate estimates of aquifer hydraulic conductivity cannot be obtained under most circumstances and it is not possible to tell whether there is a skin with a different hydraulic conductivity than that of the aquifer. Time-normalized head data for observation wells close to the slugged well (within tens of feet) are sufficiently different in shape and magnitude to allow a reasonable estimate of storativity or specific yield.

Slug test observation well type curve values with delayed response differ appreciably from type curve values with no delayed response as illustrated in

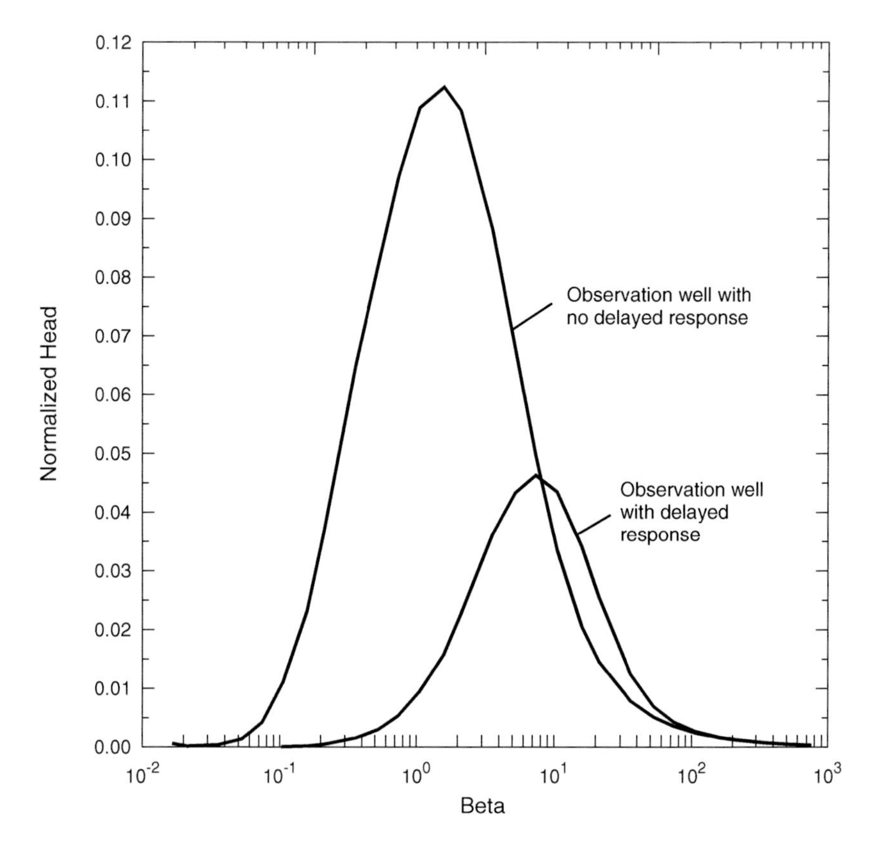

FIGURE 3.7 Graph showing observation well delayed response effects on slug test type curve values.

Figure 3.7. Type curve values with delayed response have lower peaks and are shifted to the right of type curve values without delayed response. Usually observation wells are packed off to eliminate delayed response.

HORIZONTAL ANISOTROPY

Some sedimentary and fractured aquifers are horizontally anisotropic. The horizontal hydraulic conductivity in one direction may be between 2 and 20 times or more the horizontal hydraulic conductivity in another direction. Drawdown contours around a pumped well in an anisotropic aquifer form concentric ellipses rather than circles, as they would in an isotropic aquifer. Major and minor directions of transmissivity coincide with major and minor ellipse axes. Simulation of horizontal anisotropy with numerical mathematical modeling equations can be accomplished by varying finite-difference grid cell hydraulic characteristics. Simulation is more difficult with analytical mathematical modeling equations as described below.

Aquifer test data for severe horizontal anisotropy conditions can be analyzed with the parallel boundary (strip aquifer) image well theory. Aquifer test data for less severe horizontal anisotropy can be analyzed with analytical methods derived by Hantush (1966a and 1966b) as explained in Kruseman and de Ridder (1991) and Batu (1998). These analytical methods cover the three following horizontal anisotropy conditions:

1. Principal directions of horizontal anisotropy known and ellipse of equal drawdown unknown
2. Principal directions of horizontal anisotropy unknown and ellipse of equal drawdown unknown
3. Ellipse of equal drawdown known

Time-drawdown data are matched to an appropriate family of type curves for isotropic conditions in the case of horizontal anisotropy Condition 1 or Condition 2. The effective transmissivity (T_e) is calculated with isotropic analytical mathematical modeling equations and the dimensionless drawdown and measured drawdown match point coordinates. The values of T_e calculated for all observation wells should be approximately the same. The average value of T_e is used in Equation 3.269. The effective transmissivity (T_e) is defined by the following equation (Hantush, 1966):

$$T_e = (T_x T_y)^{0.5} \tag{3.269}$$

where T_x is the transmissivity in the major direction of horizontal anisotropy and T_y is the transmissivity in the minor direction of horizontal anisotropy.

The ratio S/T_n, where S is storativity and T_n is the transmissivity in a direction that makes an angle $(\ominus + \alpha)$ with the X axis (major axis) as defined in Figure 3.8, is calculated with isotropic analytical mathematical modeling equations and the dimensionless time and measured time match point coordinates. The storativity, T_x, and T_y can be calculated provided there are one or more observation wells or piezometers on more than one ray of observation wells or piezometers. If the principal directions of horizontal anisotropy are known, two observation wells or piezometers on different rays are sufficient. If the principal directions of horizontal anisotropy are unknown, three observation wells or piezometers on different rays are required.

T_n is defined by the following equation (Hantush,1966):

$$T_n = T_x/[\cos^2(\ominus + \alpha_n) + m\sin^2(\ominus + \alpha_n)] \tag{3.270}$$

where n is the ray number and

$$m = T_x/T_y = (T_e/T_y)^2 \tag{3.271}$$

$$\alpha_1 = 0 \tag{3.272}$$

$$T_1 = T_x/(\cos^2\ominus + m\sin^2\ominus) \tag{3.273}$$

For confined nonleaky and unconfined aquifers:

$$a_n = T_1/T_n = [\cos^2(\ominus + \alpha_n) + m\sin^2(\ominus + \alpha_n)]/(\cos^2\ominus + m\sin^2\ominus) \tag{3.274}$$

where

$$a_1 = 1 \tag{3.275}$$

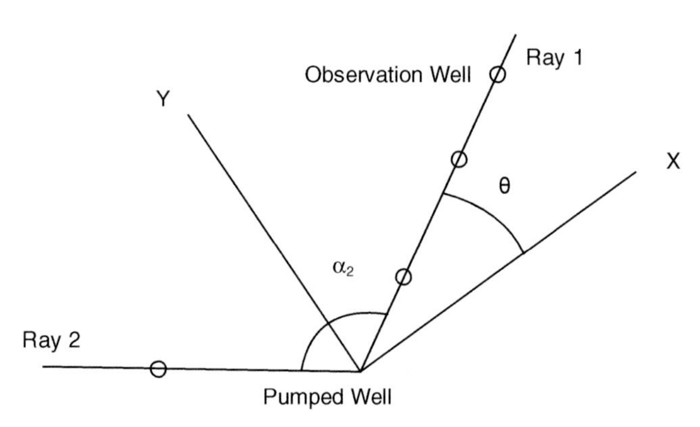

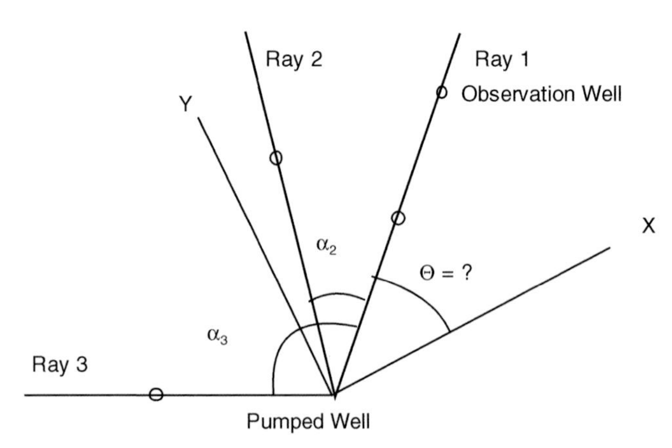

FIGURE 3.8 Horizontal anisotropy parameters.

For confined leaky aquifers:

$$a_n = 0.5[T_1/T_n + (B'_1/B'_n)^2] \tag{3.276}$$

where

$$B'_n = (T_n b'/K')^{0.5} \tag{3.277}$$

b' is the confining unit thickness and K' is the confining unit vertical hydraulic conductivity from Equation 3.271 and Equation 3.274.

$$m = (T_e/T_y)^2 = [a_n\cos^2\ominus - \cos^2(\ominus + \alpha_n)]/[\sin^2(\ominus + \alpha_n) - a_n\sin^2\ominus] \tag{3.278}$$

If the principal directions of anisotropy are not known, $\ominus$ is calculated with the following equation:

$$\tan(2\ominus) = -2\{[(a_3 - 1)\sin^2\alpha_2 - (a_2 - 1)\sin^2\alpha_3]/ [(a_3 - 1)\sin2\alpha_2 - (a_2 - 1)\sin2\alpha_3]\} \tag{3.279}$$

Equation 3.279 has two roots in the range 0 to 2π, one being the major axis of transmissivity (X axis) and the other being the minor axis of transmissivity (Y axis). If one root is δ, the other will be $\delta + \pi$. The value of $\ominus$ that makes $m > 1$ locates the major axis of anisotropy. A negative value of $\ominus$ indicates the positive X axis lies to the left of the first ray of observation wells or piezometers.

If the principal directions of anisotropy are known, a_2 is calculated with Equation 3.274 or Equation 3.276 and previously calculated ratios S/T_n. Values of $\ominus$, α_2, a_2, and T_e are substituted into Equation 3.278 to calculate m. Values of T_e and m are substituted into Equation 3.271 to calculate T_y and T_x. Values of T_x, m, $\ominus$, and α_2 are substituted into Equation 3.273 and Equation 3.274 or Equation 3.276 to calculate T_1 and T_2. Finally, values of S/T_1, S/T_2, T_1, and T_2 are used to calculate S.

If the principal directions of anisotropy are unknown, a_2 and a_3 are calculated with Equation 3.274 or Equation 3.276 and previously calculated ratios S/T_n. Values of a_2, a_3, α_2, and α_3 are substituted into Equation 3.277 to calculate $\ominus$. Values of $\ominus$, T_e, α_2, and a_2 (or α_3 and a_3) are substituted into Equation 3.278 to calculate m. Values of T_x, m, and $\ominus$ and the values of $\alpha_1 = 0$, α_2, and α_3 are substituted into Equation 3.270 to calculate T_1, T_2, and T_3. Finally, values of S/T_1, S/T_2, S/T_3, T_1, T_2, and T_3 are used to calculate S.

A Fortran computer program, Tensor2D, developed and documented by Maslia and Randolph (1987) can be used to analyze pumping test data for an anisotropic confined nonleaky aquifer. Tensor2D is based on the equation of drawdown formulated by Papadopulos (1965) for nonsteady flow in an infinite anisotropic confined nonleaky aquifer. Data for more than three observation wells or piezometers can be analyzed with a weighted least-squares optimization procedure. Several other methods for analyzing pumping test data in anisotropic aquifers are described in the literature (Way and McKee, 1982; Neuman et al., 1984; and Hsieh et al., 1985).

HORIZONTAL HETEROGENEITY

Many aquifers are horizontally heterogeneous. For example, aquifer hydraulic conductivity can progressively increase or decrease due to major depositional regimes. The aquifer stratigraphic framework can consist of several beds of different hydraulic conductivities. There can be sharp contrasts in aquifer hydraulic conductivity over limited distances. Aquifers can be trending, layered, and discontinuous. Aquifer heterogeneities can follow the complex spatial distribution of structural or sedimentologic architectural elements.

Pumping test data for horizontal heterogeneous aquifers are commonly analyzed analytically with one of the following two methods:

1. Interpret time-drawdown data for far wells to estimate effective (representative) large-scale hydraulic parameter values with homogeneous aquifer analytical equations
2. Interpret time-drawdown data for the pumped and near observation wells to estimate small-scale hydraulic parameter values with homogeneous aquifer analytical equations

Method 1 is primarily important for water supply studies. In Method 1, heterogeneities are simulated by averaging small-scale spatial hydraulic parameter variations. Method 2 is primarily important for contamination transport studies because contaminant spreading largely depends on spatial variations in hydraulic conductivity. Analytical pumping test analysis in heterogeneous aquifers is much more complex than analytical pumping test analysis in homogeneous aquifers because the sensitivity of drawdown and recovery data to heterogeneity and anisotropy varies both in space and time.

Oliver (1993) and Leven (2002) describe drawdown data sensitivity to heterogeneity during pumping tests briefly as follows:

- Drawdown is sensitive to heterogeneity within the pumping test domain, which expands in volume with time, the area of influence of heterogeneity on drawdown is elliptical in shape, and the influence is not spatially uniform within the pumping test domain.
- Drawdown is most sensitive to heterogeneity within the pumping test domain during early times.
- Drawdown sensitivity to heterogeneity depends on the contrast of hydraulic conductivity.
- Drawdown sensitivity to heterogeneity is highest close to the pumped well and the heterogeneity.

- Drawdown sensitivity to anisotropy depends on the relative locations of the principal axis of anisotropy and the pumped and observation wells.
- Hydraulic conductivity heterogeneity can either increase or decrease drawdown depending on the relative locations of the heterogeneity and the pumped and observation wells.
- Increased storativity heterogeneity decreases drawdown and decreased storativity heterogeneity increases drawdown regardless of the relative locations of the heterogeneity and pumped and observation wells.

Sanchez-Vila (1999) studied the results of analytically analyzing pumping test data for heterogeneous aquifers assuming horizontal homogeneous aquifer (Jacob's method) conditions. Briefly, Sanchez-Vila's conclusions are as follows:

- Variability in transmissivity is apparent as a variability in storativity.
- Hydraulic conductivity values calculated with late time-drawdown data for several fully penetrating observation wells tend to be uniform in space and represent the effective (geometric mean) hydraulic conductivity within the pumping test domain.
- Storativity values calculated with late time-drawdown data for several fully penetrating observation wells tend to be variable in space and are not representative by themselves.
- Real storativity can rarely be obtained by analyzing pumping test data for heterogeneous aquifers assuming horizontal homogeneous aquifer (Jacob's method) conditions.

It follows that variations in calculated storativity can be useful in diagnosing heterogeneity. Studies of data for observation wells at variable distances from the pumped well and locations on rays along and at right angles to heterogeneities also can be useful in diagnosing heterogeneity. For example, average hydraulic parameter values for the aquifer test domain can be calculated with late time-drawdown data. Theoretical distance-drawdown data can then be determined with these values and compared with measured distance-drawdown data. Deviations between theoretical and measured distance-drawdown data represent the effects of heterogeneity.

The effects of heterogeneity also can be diagnosed with pumped well specific capacity data for a selected time. As a result of heterogeneity, measured specific capacity in the pumped well, assuming complete well development and negligible well losses, will differ from theoretical specific capacity based on the average hydraulic parameter values calculated with aquifer test late time-drawdown data.

Heterogeneity also can be diagnosed with composite plots of time-drawdown data for pumped and observation wells. Evaluation of time-drawdown data observed at different locations in an aquifer may not result in one consistent set of hydraulic parameter values, which indicates that the aquifer is not homogeneous.

Several analytical time-drawdown approaches to aquifer test analysis in heterogeneous aquifers have been developed in recent decades (see Butler, 1988;

Butler, 1990; Butler, 1991; Butler and Liu, 1991; Butler and Liu, 1993; Indelman et al., 1996; Kabala, 2001; Leven, 2002; Oliver, 1993; Sanchez-Vila, 1997; Sanchez-Vila et al., 1999; Vasco et al., 2000; Yeh, 1986; Zlotnik and Ledder, 1996; and Walker and Roberts, 2003). Some of the approaches such as that of Butler (1988) are based on specified patterns of heterogeneity. It is usually assumed that transmissivity varies logarithmically in space while storativity is constant in space. Other approaches such as that of Oliver (1993) do not predefine the pattern of heterogeneity. The practical application of these methods has been limited.

Oliver (1993) presents an unsteady state flow analytical solution for transmissivity and storativity with constant pumping from a single well in a radially symmetric modest heterogeneous aquifer with transmissivity varying logarithmically in space and uniform storativity. The solution involves Frechet kernels (sensitivity coefficients) as convolution integrals in the time domain, which must be evaluated numerically. Knight and Kluitenberg (2005) present explicit analytical expressions for storativity and transmissivity Frechet kernels for both pumping and slug tests in a radially symmetric modest heterogeneous aquifer with transmissivity and storativity varying uniformly in space. The explicit analytical expressions involve Bessel functions.

Sanchez-Vila (1997) presents a steady state flow analytical solution for the effective transmissivity with constant pumping from a single well of finite radius in a heterogeneous statistically isotropic random aquifer. The solution indicates effective transmissivity is an increasing monotonic function of distance from the pumped well. Effective transmissivity rises from the harmonic mean of the point values close to the pumped well and tends asymptotically toward the geometric mean far from the pumped well.

Leven (2002) describes a consecutive multiple well approach to pumping test analysis in heterogeneous aquifers wherein several wells are located within and along the heterogeneity. The wells have small diameters and are completely developed. A constant low rate pumping test is conducted consecutively at each well. The duration of each test is short and usually 1000 sec or less. Time-drawdown data for each pumped well are plotted as semilogarithmic graphs.

Straight lines are drawn through the very early time-drawdown data when wellbore storage effects are negligible. The straight lines are extended to zero drawdown. The slopes of the straight lines and zero drawdown intercepts are used to calculate small scale values of hydraulic conductivity and storativity. These values are assigned to a pumping test domain within a radius of tens of feet of each pumped well. The time when wellbore storage effects are negligible is ascertained by noting that pumped well time-drawdown data plot as a straight line with a slope of one on a double-logarithmic time-drawdown graph during the period when wellbore storage effects are appreciable.

The slope of late time-drawdown data differs from the slope of the early time-drawdown data in heterogeneous aquifers. Large-scale effective values of hydraulic conductivity and storativity are calculated with late time-drawdown data and assigned to a pumping test domain within a radius of hundreds to thousands of feet of the pumped wells. The quality of hydraulic conductivity and storage values depends on the effectiveness of well development.

There are other approaches for analyzing pumping test data for heterogeneous aquifers. For example, information concerning aquifer heterogeneity can be obtained with large drawdown slug tests (Gonzalo Pulido, HydroQual, Inc. gpulido@hydroqual.com). Large drawdown slug tests are slug tests with large normalized heads greater than 5 m, which enhance the estimation of hydraulic parameters with data from observation wells tens of meters from the slugged well.

Vertical variations in hydraulic conductivity can be estimated with dipole flow tests as described by Butler (1998a). The dipole flow test involves a single well test in which a three-packer tool is placed in the screened (open) interval of a well. A small downhole pump moves water from one chamber of the tool to the other through the center of the middle packer, thereby setting up a circulation pattern in the adjacent aquifer. The head difference between the two chambers is used to estimate the hydraulic conductivity of near-well portions of the aquifer.

Spatial variations in hydraulic conductivity can be estimated with the direct-push method (Butler et al., 2000). This method involves performing series of slug tests in direct-push rods as the rods are driven progressively deeper into the formation. Screened intervals in the rods are only exposed to the aquifer during slug tests, thereby minimizing the amount of well development without the need of permanent wells.

Interwell variations in horizontal hydraulic conductivity can be estimated with the hydraulic tomography approach (Bohling et al., 2003; Bohling et al., 2002; Butler et al., 1999; and Yeh and Liu, 2000). This approach consists of a series of short-term pumping tests stressing different vertical aquifer intervals in networks of multilevel small-diameter sampling wells. Detailed drawdown data is obtained with miniature fiber-optic pressure sensors or air-pressure transducers. Data from all tests are analyzed simultaneously to characterize the hydraulic conductivity variation between wells.

IMAGE WELL THEORY

According to the image well theory (see Ferris et al., 1962, pp. 144–146 and Walton, 1962), a full barrier boundary is defined as a line (streamline) across which there is no flow, and it may consist of folds, faults, or relatively impervious deposits such as shale or clay. A full recharge boundary is defined as a line (equipotential) along which there is no drawdown, and it may consist of increased aquifer transmissivity or streams, lakes, and other surface water bodies hydraulically connected to the aquifer. Most full hydrogeologic boundaries are not clear-cut straight-line features but are irregular in shape and extent. However, complicated full boundaries are simulated with straight-line demarcations.

The image well theory for a full barrier boundary can be stated as follows: The effect of a full barrier boundary on the drawdown in a well, as a result of pumping from another well, is the same as though the aquifer were infinite in areal extent and a like discharging image well were located across the full barrier boundary on a perpendicular line thereto and at the same distance from the full

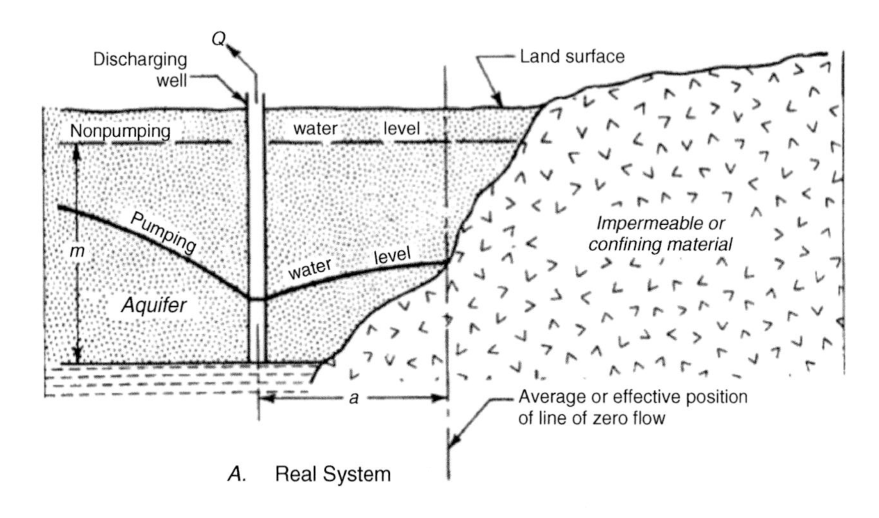

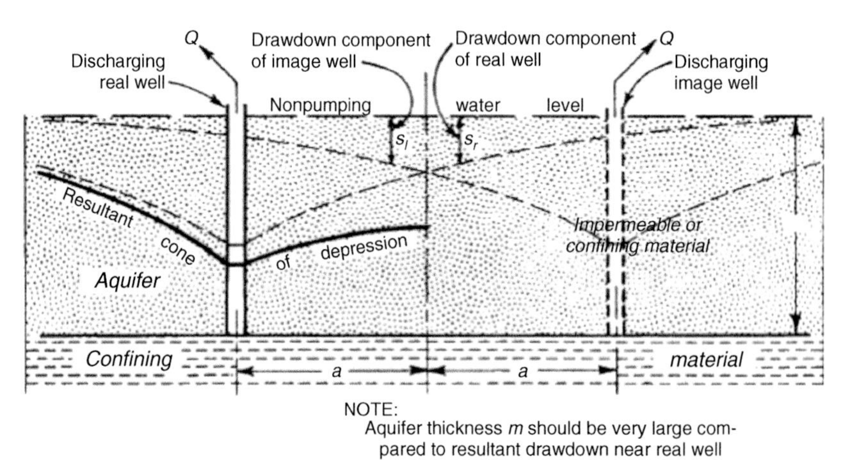

FIGURE 3.9 Image well system for barrier boundary (from Ferris et al., 1962, U.S. Geological Survey, Water-Supply Paper 1536E).

barrier boundary as the pumped well as shown in Figure 3.9. The principle is the same for a full recharge boundary except the image well is assumed to be recharging the aquifer system instead of pumping from it as shown in Figure 3.10.

The image well, like the pumped well, can have wellbore storage and can partially penetrate the aquifer. The observation well wellbore storage is influenced by the image well. Thus, the impacts of full hydrogeologic boundaries on drawdown can be simulated by use of hypothetical wells. Full boundaries are replaced by imaginary wells that produce the same disturbing effects as the boundaries. Full boundary well hydraulics problems are thereby simplified to consideration of an infinite aquifer system in which real and image wells operate simultaneously.

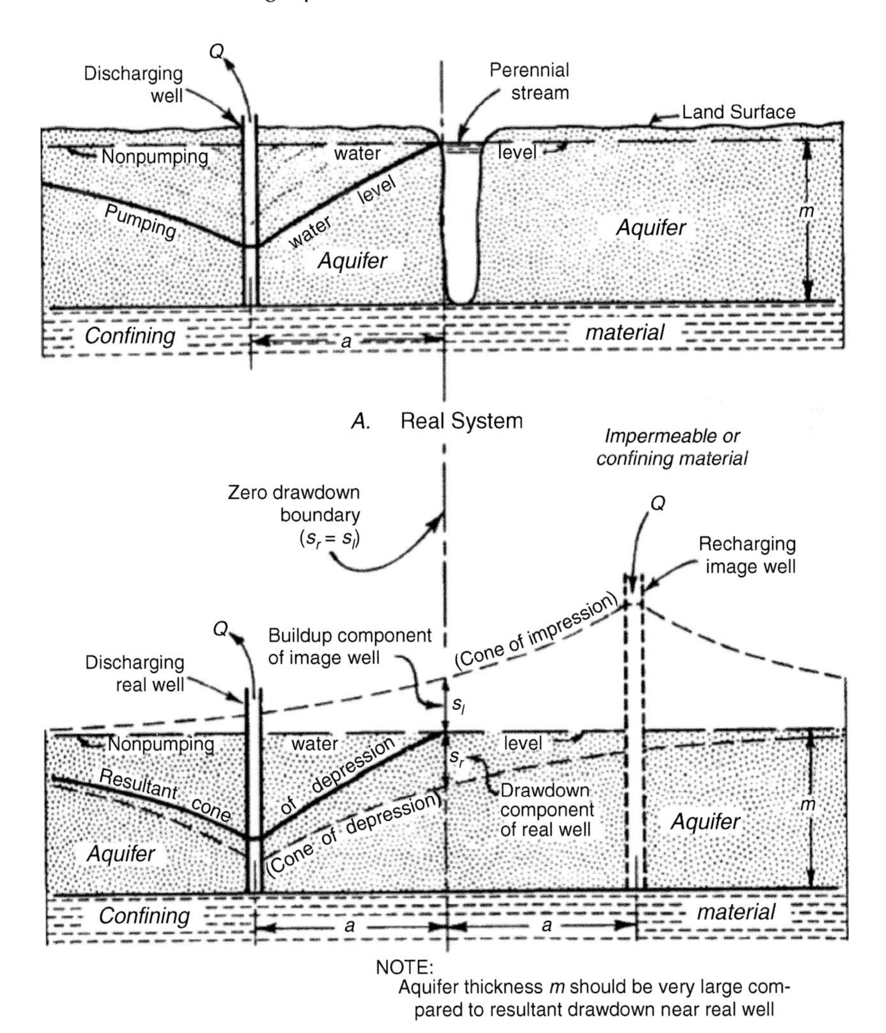

FIGURE 3.10 Image well system for recharge boundary (from Ferris et al., 1962, U.S. Geological Survey, Water-Supply Paper 1536E).

Total drawdown at any particular time is the algebraic summation of real and image well drawdown or buildup components.

A partial barrier boundary (barrier discontinuity) is defined as a line beyond which the aquifer transmissivity is much less than the aquifer transmissivity at the pumped well. A partial recharge boundary (recharge discontinuity) is defined as a line beyond which the aquifer transmissivity is much greater than the aquifer transmissivity at the pumped well. Most discontinuities are not clear-cut straight-line features but are irregular in shape and extent. However, complicated discontinuities are simulated with straight-line demarcations.

The image well theory for a discontinuity can be stated as follows: The effect of a barrier discontinuity on the drawdown in a well, as a result of pumping from another well, is the same as though the aquifer were infinite in areal extent and a discharging image well were located across the barrier discontinuity on a perpendicular line thereto and at the same distance as the pumped well. The image well discharge is a fraction of the pumped well discharge and depends on the relative aquifer transmissivities on both sides of the boundary (Muskat, 1937; Streltsova, 1988, p. 219; and McKinley and Streltsova, 1993, p. 130).

The principle is the same for a recharge discontinuity except the image well is assumed to be recharging the aquifer system instead of pumping from it. The image well, like the pumped well, has wellbore storage and can partially penetrate the aquifer. The observation well can have wellbore storage in response to the influence of the image well. Thus, the impacts of discontinuities on drawdown can be simulated by use of hypothetical wells. Discontinuities are replaced by imaginary wells that produce the same disturbing effects as the discontinuities. Discontinuity well hydraulics problems are thereby simplified to consideration of an infinite aquifer system in which real and image wells operate simultaneously. Total drawdown at any particular time is the algebraic summation of real and image well drawdown or buildup components.

The image well strength with a discontinuity can be estimated with the following equation (Muskat, 1937):

$$Q_{is} = QD_{is} \tag{3.280}$$

where

$$D_{is} = (T_p - T_d)/(T_p + T_d) \tag{3.281}$$

Q_{is} is the constant image well strength, Q is the constant pumped well discharge rate, T_p is the aquifer transmissivity between the pumped well and the discontinuity, and T_d is the aquifer transmissivity beyond the discontinuity.

Strictly speaking, equations 3.280 and 3.281 are valid only when the diffusivities (T/S) on either side of the discontinuity are equal. However, for practical purposes, the use of these equations results in reasonable discontinuity simulations. The exact equation when the diffusivities are unequal is presented by Streltsova (1988, pp. 219–220). McKinley and Streltsova (1993, p. 130) provide a monograph for analyzing a discontinuity when the diffusivities on either side of the discontinuity are unequal.

Nind (1965) presents equations for drawdown in the presence of linear discontinuities. Nonsteady state equations describing drawdown on both sides of a discontinuity in a confined nonleaky aquifer were derived by Fenske (1984).

With barrier boundaries, water levels in observation wells decline at an initial rate under the influence of the pumping well only, as if the aquifer system were infinite in areal extent. When the cone of depression of the boundary image well appreciably impacts the observation wells, the time rate of drawdown increases because the total rate of withdrawal from the aquifer system is then equal to that of the pumping well plus that of the discharging image well. Thus, the time-drawdown curve is deflected downward.

With recharge boundary conditions, water levels in observation wells decline at an initial rate under the influence of the pumping well only, as if the aquifer were infinite in areal extent. When the cone of impression of the recharging image well appreciably impacts the observation well, the time rate of drawdown changes and decreases. With a full recharge boundary, equilibrium conditions will eventually prevail and the time-drawdown curve will level off.

When a well near a stream hydraulically connected to an aquifer is pumped, the cone of depression grows until it intercepts sufficient area of the streambed and is deep enough beneath the streambed so that induced streambed infiltration balances discharge. The cone of depression may expand only partially or across and beyond the streambed depending upon the hydraulic conductivity of the streambed. The use of the image well theory to simulate induced streambed infiltration assumes that streambed partial penetration and aquifer stratification are integrated into the effective distance to the recharging image well.

Aquifers are often delimited by two or more boundaries as shown in Figure 3.11. Two converging boundaries delimit a wedge-shaped aquifer, two parallel boundaries delimit an infinite-strip aquifer, two parallel boundaries intersected at right angles by a third boundary delimit a semi-infinite strip aquifer, and four boundaries intersecting at right angles delimit a rectangular aquifer. The image well theory is applied to such cases by taking into consideration successive image well reflections on the boundaries.

A number of image wells are associated with a pair of converging boundaries. A primary image well placed across each boundary balances the impacts of the pumping well at each boundary. However, each primary image well produces an unbalanced impact at the opposite boundary. Secondary image wells must be added at appropriate positions until the impacts of the pumping and primary image wells are balanced at both boundaries. Although image well systems can be devised regardless of the wedge angle involved, simple solutions of closed image well systems are preferred. The actual aquifer wedge angle is approximated as equal to one of certain aliquot parts of 360 degrees. These approximate angles were specified by Ferris et al. (1962, p.154) as follows: If the aquifer wedge boundaries are of like character, the approximate angle must be an aliquot part of 180 degrees; if the aquifer wedge boundaries are not of like character, the approximate angle must be an aliquot part of 90 degrees; and if the pumping well is on the bisector of the wedge angle and the aquifer wedge boundaries are like in character and both barriers, the approximate angle must be an odd aliquot part of 360 degrees. Under these conditions, the exact number of image wells is equal to (360 degrees divided by the wedge angle) minus 1.

The character of each image well is the same if the aquifer wedge boundaries are of like character. If the aquifer wedge boundaries are not of like character, the character of each image well is ascertained by balancing the image well system considering each boundary separately with the following rules (Walton,

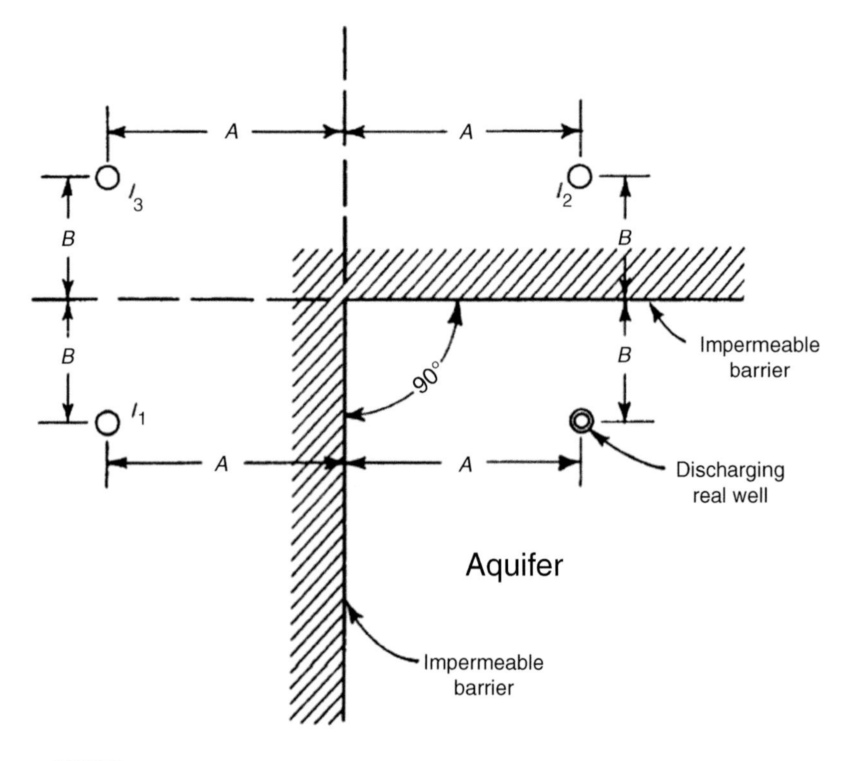

FIGURE 3.11 Image well system for wedge boundary (from Ferris et al., 1962, U.S. Geological Survey, Water-Supply Paper 1536E).

1963, pp. 20–21): A primary image well placed across a barrier boundary is discharging in character, and a primary image well placed across a recharge boundary is recharging in character; a secondary image well placed across a barrier boundary has the same character as its parent image well, and a secondary image well placed across a recharge boundary has the character opposite that of its parent image well.

Two parallel boundaries require the use of an image well system extending to infinity. Each successively added secondary image well produces a residual impact at the opposite boundary. However, in practice it is only necessary to add pairs of image wells until the next pair has negligible influence (< 0.01 ft) on the sum of all image well impacts out to that point.

If s_t is the total drawdown in an observation well at time t, s_p is the component of drawdown caused by the pumped well at time t, and s_I is the component of drawdown or buildup caused by an image well associated with a single boundary at time t, then

$$s_t = s_p + s_I \text{ for a barrier boundary} \quad (3.282)$$

$$s_t = s_p - s_I \text{ for a recharge boundary} \quad (3.283)$$

Appropriate dimensionless drawdowns are utilized to calculate s_p and s_I depending on existing aquifer conditions. For example, with confined nonleaky aquifer conditions:

$$s_o = QW(u)/(4\pi T) + Q_I W(u_I)/(4\pi T) \text{ for a barrier boundary} \quad (3.284)$$

$$s_o = QW(u)/(4\pi T) - Q_I W(u_I)/(4\pi T) \text{ for a recharge boundary} \quad (3.285)$$

where

$$u = r^2 S/(4Tt) \quad (3.286)$$

$$u = r_i^2/(4Tt) \quad (3.287)$$

Q is the pumped well discharge rate, T is the aquifer transmissivity, r is the distance from the pumped well to the observation well, S is the aquifer storage, t is the elapsed time, r_i is the distance from the image well to the observation well, $W(u)$ is the dimensionless drawdown for the pumped well and $W(u_I)$ is the dimensionless drawdown or recovery for the boundary image well.

NUMERICAL MATHEMATICAL MODELING EQUATIONS

Numerical groundwater flow mathematical modeling equations can generate dimensionless or dimensional time-drawdown values for simplistic conceptual models equally as well as analytical models. In addition, numerical models can generate dimensionless and dimensional time-drawdown values for complex conceptual models. Both numerical and analytical models can simulate a homogeneous medium with a uniform thickness, confining unit storativity, delayed gravity drainage under unconfined aquifer conditions, wellbore storage, partially penetrating wells, boundaries as straight-line demarcations, single aquifer and confining unit, and a pumped well open to only one aquifer. Numerical models can simulate a heterogeneous aquifer with a nonuniform thickness, irregular boundaries, multiple aquifer and confining unit layers, and a pumped well open to several aquifers.

Numerical models are based on the following partial-differential equation describing the three-dimensional movement of groundwater of constant density through heterogeneous and anisotropic porous earth material under nonequilibrium conditions (McDonald and Harbaugh, 1988, p. 2-1):

$$\partial/\partial_x(K_{xx}\partial h/\partial x) + \partial/\partial_Y(K_{YY}\partial h/\partial y) + \partial/\partial_z(K_{zz}\partial h/\partial z) - W = S_s\partial h/\partial t \quad (3.288)$$

where K_{xx}, K_{YY}, and K_{zz} are values of hydraulic conductivity along the *x*-, *y*-, and *z*-coordinate axes, which are assumed to be parallel to the major axes of hydraulic conductivity, *h* is the potentiometric head, *W* is a volumetric flux per unit volume and represents sources or sinks of water, S_s is the specific storage of the porous material, and *t* is time. The principal axes of hydraulic conductivity are assumed to be aligned with the coordinate directions. S_s, K_{xx}, K_{YY}, and K_{zz} may be functions of space [$S_s = S_s(x, y, z)$, $K_{xx} = K_{xx}(x, y, z)$, etc.] and *W* may be a function of space and time [$W = W(x, y, z, t)$].

The mathematical modeling equations describing groundwater flow consist of the partial-differential equation together with specification of flow or head conditions at the boundaries of an aquifer system and specification of initial head conditions. Numerical mathematical modeling equations commonly utilize the finite-difference approximation method (McDonald and Harbaugh, 1988; Anderson and Woessner, 1992).

FINITE-DIFFERENCE APPROXIMATION METHOD

In the finite-difference approximation method, the continuous system described by partial-differential Equation 3.288 is replaced by a finite set of discrete points in space and time. Partial derivatives are replaced by terms calculated from the differences in head values at these points. Systems of simultaneous linear algebraic difference equations are generated and expressed as matrix equations and iterative numerical methods are used to solve the matrix equations. The solution of matrix equations leads to values of head at specific points and times.

The aquifer test conceptual model is replaced by a discretized grid of nodes centered at the pumped well and associated finite-difference cells (blocks) simulating one or more aquifer layers. Delayed gravity drainage under unconfined aquifer conditions is simulated with 10 or more confined layers and 1 unconfined layer. Parameter values are assigned to grid cell groups and boundary conditions are simulated along or within grid cell borders. Initial conditions are simulated by assigning the same head to all grid nodes. Aquifer test time and the pumping rate are discretized into small blocks of variable lengths to simulate wellbore storage.

The computer program MODFLOW, developed by the U.S. Geological Survey, is a prime example of the implementation of the finite-difference approximation method. MODFLOW requires several input files and produces several output files. Detailed instructions for the preparation of input files are provided in MODFLOW documentations.

SOFTWARE SELECTION

A large variety of public domain and commercial aquifer test analysis software is available with a broad range of sophistication. Primary analytical and numerical software usually contains code to read input data files, code to calculate either or both dimensionless and dimensional time-drawdown values based on file input data (calculation engine), and code to generate output files for use with external paper graphs or external word processor, spreadsheet, database, graphics software, and automatic parameter estimation software. Primary software usually is distributed by governmental agencies or universities. Sophisticated analytical and numerical software contains code for interactive computer screen input (preprocessor), a calculation engine, internal automatic parameter estimation code, and code to display calculation results on the computer screen or with a printer (postprocessor). Sophisticated software can also contain integrated word processor, spreadsheet, database, and graphics capabilities for seamless analysis. Sophisticated software is usually distributed commercially. Less sophisticated software is included with some aquifer test analysis books.

The U.S. Geological Survey distributes the fully documented primary analytical software WTAQ described by Barlow and Moench (1999). WTAQ is written in Fortran and contains state-of-the-art code for calculating analytical Stehfest aquifer test model dimensionless or dimensional time-drawdown values with confined nonleaky or unconfined (water table) aquifer conditions.

The U.S. Geological Survey also distributes several versions of the fully documented primary numerical software MODFLOW described by McDonald and Harbaugh (1988). MODFLOW has become an international standard.

The WTAQ and MODFLOW software and documentation are available free of charge at water.usgs.gov/nrp/gwsoftware/.

Most analytical integral and numerical model commercial software can be purchased from the Scientific Software Group. P.O. Box 708188, Sandy, Utah 84070, (801) 208–3011 or at www.scisoftware.com.

Detailed information concerning commercial and free software can be obtained at:

www.groundwatermodels.com
www.flowpath.com
www.AQTESOLV.com
www.Aquifer-Test.com
typhoon.mines.edu/software/igwmcsoft/
www.rockware.com

The universal automatic parameter estimation software PEST and documentation and PEST Groundwater Data Utilities can be obtained free of charge at www.sspa.com/pest/.

The universal automatic parameter estimation software UCODE and documentation can be obtained free of charge at www.typhoon.mines.edu/software/igwmcsoft/.

The following books contain analytical integral aquifer test model software:

Dawson, K.J. and J.D. Istok. 1991. *Aquifer Testing: Design and Analysis of Pumping and Slug Tests*. Lewis Publishers, Boca Raton, FL at www.crcpress.com.

Halford, K.J. and E.L. Kuniansky. 2002. Spreadsheets for the Analysis of Aquifer-Test and Slug-Test Data. U.S. Geological Survey Open-File Report 02-197 at water.usgs.gov/pubs/of/.

Hall, Phil. 1996. *Water Well and Aquifer Test Analysis*. Water Resources Publications, LLC. Highlands Ranch, CO at www.wrpllc.com.

Boonstra, J. and R.A.L. Kselik. 2002. SATEM: Software for Aquifer Test Evaluation. Publication 57. International Institute for Land Reclamation and Improvement. The Netherlands at www.alterra-research.nl/pls.

Batu, Vedat. 1998. *Aquifer Hydraulics: A Comprehensive Guide to Hydrogeologic Data Analysis*. John Wiley Interscience Publications. Somerset, NJ.

Analytical Stehfest aquifer test model software is contained in the following book: Walton, W.C. 1996. *Aquifer Test Analysis with Windows Software*. Lewis Publishers, Boca Raton, FL at www.crcpress.com.

Both Fortran and Mathematica macros are provided in the following book: Cheng, A.H.-D. 2000. *Multilayered Aquifer Systems — Fundamentals and Applications*. Marcel Dekker, Inc., New York at www.amazon.com.

The following book gives instructions for ordering slug test analytical Stehfest aquifer test model software: Butler, J.J. 1998. *The Design, Performance, and Analysis of Slug Tests*. Lewis Publishers. Boca Raton, FL at www.crcpress.com.

The following book contains MODFLOW pre- and postprocessor software (PMWIN): Chiang, Wen-Hsing and Wolfgang Kinzelbach. 2003. 3D-*Groundwater Modeling with PMWIN*. Springer-Verlag, New York at www.uovs.ac.za/faculties.

The following book gives instructions for obtaining numerical pumping test software: Lebbe, L.C. 1999. *Hydraulic Parameter Identification — Generalized Interpretation Method for Single and Multiple Pumping Tests*. Springer-Verlag, New York at allserv.ugent.be/~luclebbe.

4 External Influence Data Adjustment

The third step in aquifer test modeling is the review of the mathematical modeling equation assumptions and the adjustment of data for any departures (herein called external influences) from the assumptions. Erroneous conclusions about aquifer parameter values and boundaries can be reached if the impacts of any external influences are not removed before aquifer test data are analyzed with mathematical modeling equations. External influence fluctuations include those caused by groundwater flow through the aquifer test domain prior to the test (antecedent trend), atmospheric pressure changes, surface water (tidal, lake, or stream) stage changes, earth tides, earthquakes, applications of heavy loads (railroad trains or trucks), evapotranspiration, recharge from rainfall, and nearby pumped well pumping rate changes.

Data measurements prior to, during, and after the pumping or slug test are required for external influence adjustment. Ideally, the pumping pretest and posttest measurement period lengths should be 5 days and the pumping pretest measurement period frequency should be 1 h (Spane, 2002). External influence data adjustments are based on data reference baselines that usually extend horizontally through the time immediately prior to test initiation and vertically through prominent external influence peaks and troughs. External influence data adjustment involves interpolation and extrapolation that can be performed with data fitting computer programs in which a line or curve is fitted to data from past times and extended to estimate data for future times with linear or curvilinear polynomial approximation (regression) equations.

Usually, groundwater levels are either rising or declining due to groundwater flow prior to the test initiation instead of being constant as assumed in mathematical modeling equation assumptions. Linear or curvilinear regression of pretest data can be used to define the antecedent trend. A line or curve is fitted to the pretest time-groundwater level data usually with the least-squares method. The reference time is the groundwater level measurement initiation time. The equation of the best line or curve defines the antecedent trend. Another method for defining the antecedent trend is to adjust pretest groundwater data for any external influences and to fit a line or curve to the adjusted pretest data.

The antecedent trend is extrapolated through the test period. Differences between the extrapolated antecedent trend and the groundwater level measured immediately prior to test initiation define antecedent trend adjustments. The adjustments are either subtracted from or added to measured drawdowns during the test depending on whether the antecedent trend is rising or declining.

Suppose mathematical modeling equations assume a single pumped or slugged well. If a nearby production well is operating in addition to the aquifer test pumped well or slugged well, measured drawdowns or normalized heads are affected by the combined impacts of two wells instead of one. In this case, drawdowns or normalized heads due to any changes in the operation of the nearby production well estimated with mathematical modeling equations, the principle of superposition, and a conceptual model are subtracted from the measured drawdowns or normalized heads. According to the principle of superposition, two or more drawdown or normalized head solutions, each for a given set of aquifer and well conditions, can be summed algebraically to obtain a solution for the combined condition.

Suppose the mathematical modeling equations assume the pumping rate is constant. If the pumping rate is not constant, a best-fit constant rate is selected and drawdowns or buildups due to departures from the best-fit constant rate estimated with mathematical modeling equations, the principle of superposition, and the conceptual model are subtracted or added from the measured drawdowns.

Mathematical modeling equations assume water levels are constant throughout the aquifer prior to pumping. Suppose groundwater levels are declining prior to pumping a nearby well. In this case, the declining groundwater level trend prior to pumping (herein called the antecedent trend) is extrapolated through the pumping period. A curve or straight line is drawn through the groundwater level just before pumping started. Differences between the extrapolated antecedent trend at measurement times are calculated and subtracted or added from measured drawdowns.

Mathematical modeling equations also assume the antecedent trend does not change during the pumping period due to barometric or surface water changes. Groundwater levels can respond to temporal variations in atmospheric pressure (see Jacob, 1940; Ferris et al., 1962; Clark, 1967; Weeks, 1979; Davis and Rasmussen, 1993; and Rasmussen and Crawford, 1997). Atmospheric pressure changes may exceed 1 in. of mercury (per 1.13 ft of water). There is an inverse relationship between groundwater level and atmospheric pressure changes. Groundwater levels decline with an increase in atmospheric pressure and rise with a decrease in atmospheric pressure. The ratio of the change in groundwater level to a corresponding change in atmospheric pressure is known as the barometric efficiency. Temporal barometric efficiencies can be constant or variable.

Groundwater level response depends on whether the aquifer is confined or unconfined and whether there is wellbore storage or wellbore skin. Rasmussen and Crawford (1997) identified three conceptual models that describe groundwater level response to atmospheric pressure changes:

1. Instantaneous groundwater level response within confined aquifers
2. Delayed groundwater level response within unconfined aquifers (because of the delayed transmission of barometric pressure through the vadose zone)
3. Delayed groundwater level response associated with well characteristics (because of wellbore storage and wellbore skin)

In confined aquifers, groundwater level response depends on the degree of aquifer confinement, the rigidity of the aquifer matrix, and the specific weight of groundwater. Groundwater level change represents only that portion of the atmospheric pressure change not borne by the aquifer matrix. High barometric efficiencies reflect high strength and rigid aquifers, and low barometric efficiencies indicate highly compressible aquifers. Barometric efficiencies commonly range from 0.3 to 0.7. An inelastic aquifer could have a barometric efficiency of 1.0.

Barometric efficiency with confined aquifers and wells having no wellbore storage and skin is constant and does not vary over time. Barometric efficiency with confined aquifers and wells having wellbore storage or skin is variable, increases over time, and slowly approaches that with wells having no wellbore storage and skin as shown in Figure 4.1. Wellbore storage effects are negligible with transmissivities > 100 ft^2/day, which increase with increases in wellbore radius and decrease with increases in storativity. Commonly, wellbore storage effects shortly after an atmospheric pressure change occurs are a few hundreds of a foot or less and soon become negligible.

In unconfined aquifers with well-screen completion below the water table, atmospheric pressure changes are transmitted instantaneously at the well; however, there is a delayed response at the water table because air must move into or out of the overlying vadose zone to transmit the change in atmospheric pressure. Well-screen completion across the water table allows atmospheric pressure fluctuations to be directly imposed to the water table through the well. The rate of air movement within the vadose zone varies with its vertical pneumatic diffusivity

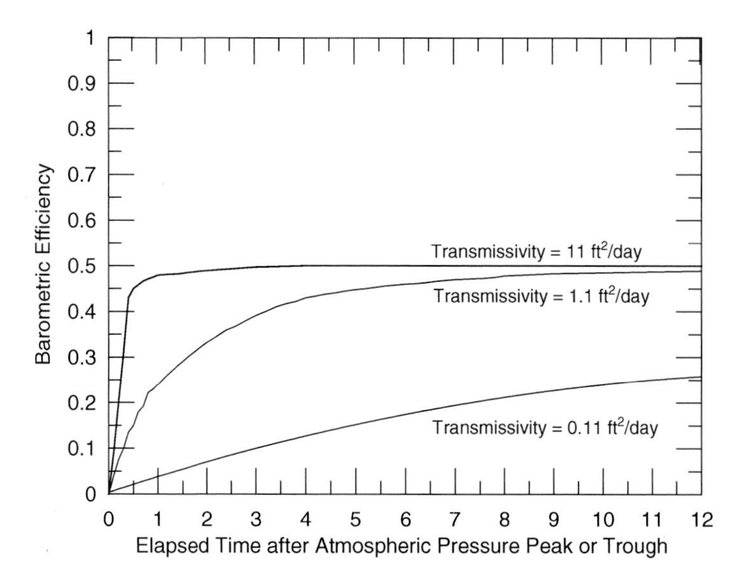

FIGURE 4.1 Typical time-barometric efficiency curves for confined aquifers (after Spane, 1999, PNNL-13078, Pacific Northwest National Laboratory, Richland, Washington).

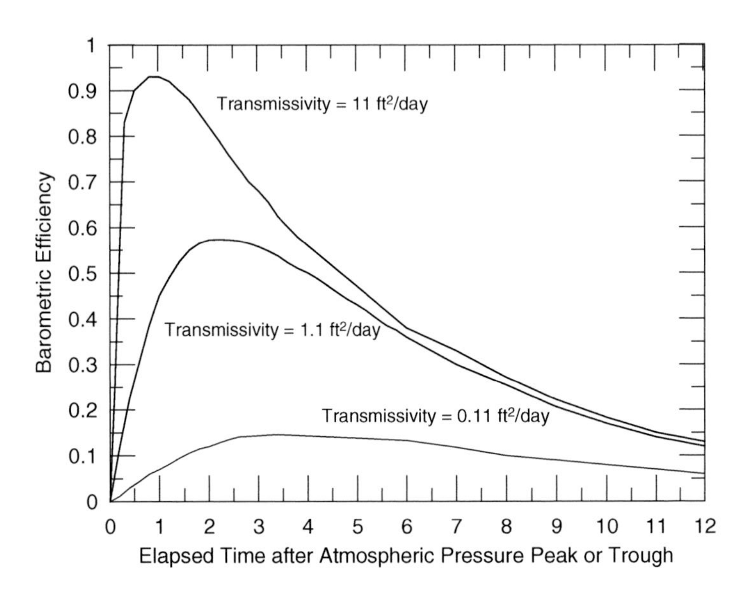

FIGURE 4.2 Typical time-barometric efficiency curves for unconfined aquifers (after Spane, 1999, PNNL-13078, Pacific Northwest National Laboratory, Richland, Washington).

(Weeks, 1979), which in turn is a function of the vadose zone vertical permeability, moisture content, and compressibility of contained gas. For vadose zones exhibiting low pneumatic diffusivities or significant thickness, unconfined aquifer wells will exhibit major groundwater level response to atmospheric pressure changes.

Barometric efficiency with unconfined aquifers and wells having no wellbore storage and skin is variable and decreases over time due to delayed transmission of atmospheric pressure through the vadose zone. Barometric efficiency with unconfined aquifers and wells having wellbore storage and skin is variable and is influenced by increases over time due to wellbore storage or skin as well as decreases over time due to delayed transmission of barometric pressure through the vadose zone as shown in Figure 4.2. Wellbore storage effects are negligible with transmissivities > 100 ft^2/day, increase with increases in wellbore radius, and increase with decreases in storativity. Commonly, wellbore storage effects shortly after an atmospheric pressure change occurs are a few hundreds of a foot or less and soon become negligible. Barometric efficiencies with unconfined aquifers can decrease from 0.8 to 0.1 or less during a 24-h period.

Groundwater level data are corrected for atmospheric pressure changes using the following equation (Ferris et al., 1962, p. 85):

$$\Delta W = (BE)\Delta B \tag{4.1}$$

where ΔW is the water level change in a well, BE is the barometric efficiency, and ΔB is the corresponding atmospheric pressure change expressed in feet of water.

Differences between the atmospheric pressure measured immediately prior to test initiation and the atmospheric pressure at any time during the test multiplied by the barometric efficiency define atmospheric pressure adjustments. The adjustments are either subtracted from or added to measured drawdowns during the test depending on whether the atmospheric pressure is rising or declining.

CONFINED AQUIFER BAROMETRIC EFFICIENCY

The order of magnitude of barometric efficiency for confined granular aquifers with no wellbore storage or skin can be estimated with the following equation (see Ferris et al., 1962, p. 90):

$$BE = (\gamma \theta b \beta)/S \tag{4.2}$$

where γ is the specific weight of groundwater at a stated reference temperature (usually assumed to be equal to 62.4 lb/ft^3), θ is the aquifer porosity, β is the bulk modulus of compression of groundwater (reciprocal of the bulk modulus of elasticity, usually assumed to be equal to .0000033 in^2/lb), and BE is the barometric efficiency.

The order of magnitude of aquifer porosity values can be estimated with Table 4.1. Storativity usually ranges from 10^{-5} to 10^{-3} and is about 10^{-6} per foot of aquifer thickness (Lohman, 1972, p. 8).

Clark (1967) describes the following method for calculating barometric efficiency when the aquifer is confined, wells have no wellbore storage and skin, there is instantaneous transmission of atmospheric pressure effects, and groundwater level changes during the time of interest are due to changes in atmospheric pressure as well as other influences such as an antecedent trend. Groundwater level and atmospheric pressure vs. time arithmetic graphs are divided into small (hourly) increments of time. The method assigns a positive sign to groundwater level changes when the groundwater level is rising and a positive sign to atmospheric pressure changes when the atmospheric pressure is decreasing. Summations of groundwater level changes and corresponding atmospheric pressure changes are calculated and tabulated for each increment of time using the following rules (Davis and Rasmussen, 1993):

- When the atmospheric pressure change is zero, neglect the corresponding value of groundwater level change in calculating summations of groundwater level changes.
- When the groundwater level change and the atmospheric pressure change have like signs, add the absolute value of the groundwater level change in calculating the summation of groundwater level changes.
- When the groundwater change and the atmospheric pressure change have unlike signs, subtract the absolute value of the groundwater level change in calculating the summation of groundwater level changes.
- Summation of atmospheric pressure change is the sum of the absolute values of atmospheric pressure change.

TABLE 4.1
Typical Porosity Ranges

Deposit	Porosity, Dimensionless
Volcanic, pumice	0.80–0.90
Peat	0.60–0.80
Soils	0.50–0.60
Silt	0.35–0.60
Clay	0.35–0.60
Loess	0.40–0.55
Sand, dune	0.35–0.55
Sand, fine	0.20–0.55
Sand, coarse	0.15–0.35
Gravel, coarse	0.25–0.35
Gravel, medium	0.15–0.25
Sand and gravel	0.20–0.35
Till	0.25–0.45
Siltstone	0.25–0.40
Sanstone	0.05–0.50
Volcanic, vesicular	0.10–0.50
Volcanic, tuff	0.10–0.40
Limestone	0.05–0.50
Schist	0.05–0.50
Shale, weathered	0.30–0.50
Basalt	0.01–0.35
Shale, at depth	0.01–0.10
Volcanic, dense	0.01–0.10
Igneous, fractured	0.01–0.10
Dolomite	0.34–0.60
Chalk	0.023–0.20
Salt, bedded	0.005–0.03
Bedded salt	0.001–0.005
Granite	0.0005–0.08
Igneous, unfractured	0.0001–0.01
Anhydrate	0.005–0.05

Based on data in Morris and Johnson, 1967, U.S. Geological Survey. Water-Supply Paper 1839-D; Walton, 1991, *Principles of Ground Water Engineering*, Lewis Publishers, Inc., pp. 414–416; Spitz and Moreno, 1996, *A Practical Guide to Ground Water and Solute Transport Modeling*, John Wiley & Sons, Inc., pp. 341–354.

The groundwater level change can be positive or negative depending on the corresponding value of atmospheric pressure change.

Incremental summations are plotted on an arithmetic graph with groundwater level incremental summations on the *y* axis and atmospheric pressure change incremental summations on the x axis. A straight line is fitted to the graph. The barometric efficiency is calculated from the slope of the graph.

Usually, the Clark methods can be applied with little error when the aquifer is confined and wells have wellbore storage.

UNCONFINED AQUIFER BAROMETRIC EFFICIENCY

Spane (2002) describes the use of two methods for calculating barometric efficiency when the aquifer is unconfined and temporal barometric efficiency is variable:

1. Multiple-regression deconvolution technique (Rasmussen and Crawford, 1997)
2. Vadose zone model (WBAR program of Weeks, 1979)

The vadose zone model requires minimal baseline data and physical system properties (e.g., pneumatic diffusivity, vadose zone thickness, background water-table trend) can be determined directly from the analysis but do not account for wellbore storage or skin and situations where the water table occurs with the well-screen section. The multiple-regression method requires longer baseline data periods to be effective and quantitative characterization of the physical system properties controlling the barometric response cannot be directly determined. Several days of baseline data may be required with a measurement frequency of every hour.

In the multiple-regression deconvolution technique with atmospheric pressure and groundwater level changes recorded every hour, the relationship between groundwater level and atmospheric pressure change is as follows (Spane, 2002):

$$\Delta h_w = X_0\,\Delta h_{ai} + X_1\Delta h_{ai-1} + X_2\Delta h_{ai-2} + \ldots + X_n\Delta h_{ai-n} \tag{4.3}$$

where Δh_w is the groundwater level change over the last hour, Δh_{ai} is the atmospheric pressure change over the last hour, Δh_{ai-1} is the atmospheric pressure change from 2 to 1 h previous, Δh_{ai-n} is the atmospheric pressure change from n h to $(n-1)$ h previous, $X_0 \ldots X_n$ are the regression coefficients corresponding to time lags of 0 to n h, and n is the number of hours that lagged atmospheric pressure effects are apparent.

The following alternative method can be used to calculate barometric efficiency when the aquifer is unconfined and temporal barometric efficiency is variable. The pretest atmospheric pressure and groundwater level records are examined and prominent peaks and troughs are selected for analysis. Reference times for each atmospheric pressure peak and trough are recorded. Incremental atmospheric pressure and related groundwater level changes at 1 h intervals after the reference time for each peak and trough are determined. The number of time intervals depends on the length of time between adjoining atmospheric pressure peaks and troughs.

The barometric efficiency for each time interval is calculated as the ratio of the incremental groundwater level change and the related incremental atmospheric pressure change. The elapsed times from peak or trough reference times to each time interval are also determined. The best-fit curve through the elapsed time (lag time) vs. barometric efficiency data is determined. The equation of the curve and atmospheric pressure peak and trough data during the aquifer test are used to correct measured drawdowns for atmospheric pressure fluctuations.

BAROMETRIC EFFICIENCY CALIBRATION

Barometric efficiencies are calibrated with pretest data before they are used to correct measured drawdowns during an aquifer test. The purpose of the calibration process is to minimize or eliminate pretest groundwater level fluctuations resulting from atmospheric pressure changes. The pretest antecedent trend defined with linear or curvilinear regression is compared with the antecedent trend defined by correcting measured pretest groundwater level data with calculated barometric efficiencies. If the comparison is favorable, the barometric efficiencies are declared to be valid. Otherwise, barometric efficiencies are adjusted and the process is repeated.

STREAM STAGE ADJUSTMENTS

Sinusoidal changes in groundwater levels will occur in response to correlative changes in stream stage (Ferris, 1951; Ferris et al., 1962). The amplitude of groundwater level fluctuations decreases with distance from the stream and the time lag of a given groundwater level peak or trough increases with distance from the stream. The amplitude and time lag depend in part on the aquifer transmissivity and storativity. Adjustments for any surface water stage changes that occur during the pumping period are calculated by comparing surface water stage changes with corresponding groundwater level changes measured prior to the test within the aquifer test domain or during pumping outside the aquifer test domain. Surface water stage change adjustments are made by first calculating the surface water efficiency, *SE,* as described by the following equation (Ferris et al., 1962, p. 85):

$$SE = (\Delta W/\Delta S) \tag{4.4}$$

where ΔW is the water level change in a well and ΔS is the corresponding surface water stage change.

Surface water efficiency is variable over time due to delayed transmission of stream stage changes through the aquifer and observation wellbore storage and skin. The following method can be used to calculate surface water efficiency. The pretest stream stage and groundwater level records are examined and prominent peaks and troughs are selected for analysis. Reference times for each surface water stage peak and trough are recorded. Incremental stream stage and related groundwater level changes at 1 h intervals after the reference time for each peak and trough are determined. The number of time intervals depends on the length of time between adjoining stream stage peaks and troughs.

The surface water efficiency for each time interval is calculated as the ratio of the incremental groundwater level change and the related incremental stream stage change. The elapsed times from peak or trough reference time to each time interval are also determined. The best-fit curve through the elapsed time (lag time) vs. surface water efficiency data is determined. The equation of the curve and stream stage peak and trough data during the aquifer test are used to correct measured drawdowns for surface water stage fluctuations.

Groundwater level data are adjusted for surface water stage changes using the following equation (Ferris et al., 1962, p. 85):

$$\Delta W = (SE)\Delta S \tag{4.5}$$

where ΔW is the water level change in a well, SE is the surface water efficiency, and ΔS is the corresponding surface water stage change.

TIDAL FLUCTUATION ADJUSTMENTS

Changes in groundwater levels near coastal waters will occur in response to correlative tidal fluctuations (see Ferris et al., 1962; Erskine, 1991; Serfes, 1991; Milham and Howes, 1995; and Trefry and Johnson, 1998). The amplitude of groundwater level fluctuations decreases with distance from the coast and the time lag of a given groundwater level peak or trough increases with distance from the coast. The amplitude and time lag depend in part on the aquifer transmissivity and storativity. Adjustments for any tidal fluctuations that occur during the pumping period are calculated by comparing tidal fluctuations with corresponding groundwater level changes measured prior to the test within the aquifer test domain or during pumping outside the aquifer test domain. Tidal fluctuation adjustments are made by first calculating the tidal efficiency, TE, as described by the following equation (Ferris et al., 1962, p. 85):

$$TE = (\Delta W/\Delta S) \tag{4.6}$$

where ΔW is the water level change in a well and ΔS is the corresponding tidal fluctuation.

Tidal efficiency is variable over time due to delayed transmission of tidal fluctuations through the aquifer and observation wellbore storage and skin. The following method can be used to calculate tidal efficiency. The pretest tidal stage and groundwater level records are examined and prominent peaks and troughs are selected for analysis. Reference times for each tidal stage peak and trough are recorded. Incremental tidal stage and related groundwater level changes at

1 h intervals after the reference time for each peak and trough are determined. The number of time intervals depends on the length of time between adjoining tidal stage peaks and troughs.

The tidal efficiency for each time interval is calculated as the ratio of the incremental groundwater level change and the related incremental tidal stage change. The elapsed times from peak or trough reference time to each time interval are also determined. The best-fit curve through the elapsed time (lag time) vs. tidal efficiency data is determined. The equation of the curve and tidal stage peak and trough data during the aquifer test are used to correct measured drawdowns for tidal stage fluctuations.

Groundwater level data are adjusted for tidal stage changes using the following equation (Ferris et al., 1962, p. 85):

$$\Delta W = (TE)\Delta S \tag{4.7}$$

where ΔW is the water level change in a well, TE is the tidal efficiency, and ΔS is the corresponding tidal stage change.

DEWATERING ADJUSTMENTS

Aquifer test mathematical modeling equations assume that drawdown in unconfined aquifers is negligible in comparison to the initial saturated thickness. Gravity drainage of interstices during a pumping test may decrease the saturated thickness of the aquifer and, therefore, aquifer transmissivity. If this is case, measured drawdowns are adjusted for the effects of aquifer dewatering with the following equation (Jacob, 1944):

$$s_{ad} = s_m - s_m^2/(2b) \tag{4.8}$$

where s_{ad} is the adjusted drawdown, s_m is the measured drawdown, and b is the initial aquifer thickness.

This equation is strictly applicable to late drawdown data and not to early and intermediate data (Neuman, 1975a, pp. 334–335).

5 Data Analysis

The next step in aquifer test modeling is the analysis of adjusted aquifer test data using the previously defined conceptual model, selected mathematical modeling equations, and selected software. Aquifer test data analysis procedures are:

- Format selection
- Technique selection
- Well function or drawdown calculation
- Calibration

The traditional approach to aquifer test data analysis involves the use of analytical integral mathematical modeling equation well functions (dimensionless drawdown values or normalized head values at dimensionless times). The following pdf documents illustrate the traditional approach to aquifer test analysis:

Hiergesell, R.A., M.K. Harris, W.E. Jones, and G.P. Flach. 2000. Results of Aquifer Tests Performed near R-Area Savannah River Site (U). WSRC-TR-2000–00180. Westinghouse Savannah River Company, Savanna River Site, Aiken, SC 29808 at sti.srs.gov/fulltext/tr2000180/ tr2000180.pdf.

Greene, E.A. 1993. Hydraulic Properties of the Madison Aquifer System in the Western Rapid City Area, South Dakota. U.S. Geological Survey Water- Resources Investigations Report 93–4008 at water.usgs.gov/pubs.

Warner, D. 1997. Hydrogeologic Evaluation of the Upper Floridan Aquifer in the Southwest Albany Area, Georgia. U.S. Geological Survey Open-File Report 97-4129 at ga.water.usgs.gov/pubs/wrir/wrir97-4129/pdf/ wrir97-4129.pdf.

Goode, D.J. and L.A. Seneor. 1998. Review of Aquifer Test Results for the Lansdale Area, Montgomery County, Pennsylvania. U.S. Geological Survey Open-File Report 98-294 at pubs.water.usgs.gov/ofr98294.

Aquifer test analysis with analytical Stehfest algorithm mathematical modeling equation well functions is illustrated in the following pdf document: Moench, A.F., S.P. Garabedian, and D.R. LeBlanc. 2001. Estimation of Hydraulic Parameters from an Unconfined Aquifer Test Conducted in a Glacial Outwash Deposit, Cape Cod, Massachusetts. U.S. Geological Survey. Professional Paper 1629 at water.usgs.gov/pubs.

FORMAT SELECTION

There are two aquifer test analysis formats:

1. Dimensionless
2. Dimensional

Dimensionless format refers to the interactive fitting (calibration) of measured (adjusted) time-drawdown or time-normalized head values to calculated dimensionless time-drawdown or time-normalized head double-logarithmic or semilogarithmic graphs (type curve or straight line matching). The graphs are usually computer screen displays. *Dimensional format* refers to the interactive or automated calibration of calculated and measured time-drawdown or time-normalized head values.

TECHNIQUE SELECTION

There are four aquifer test analysis techniques:

1. Single plot type curve matching with interactive calibration
2. Single plot straight line matching with interactive calibration
3. Composite plot type curve matching with interactive calibration
4. Composite plot with automatic parameter estimation

Single plot type curve or straight line with interactive calibration techniques usually precede and guide composite plot type curve matching and composite plot with automatic parameter estimation techniques. Accentuation of early dimensionless and measured pumping test time-drawdown data in double-logarithmic graphs facilitates the analysis of wellbore storage, well partial penetration, delayed drainage at the water table under unconfined aquifer conditions, and aquifer boundary impacts. Arithmetic values of measured drawdown plotted on semilogarithmic paper against the logarithms of elapsed time describe a straight line except during early elapsed times. Single plot type curve matching with interactive calibration is best suited for the analysis of individual observation well data, whereas, single plot straight line matching with interactive calibration is best suited for the analysis of pumped or slugged well data.

SINGLE PLOT TYPE CURVE MATCHING WITH INTERACTIVE CALIBRATION

To illustrate the single plot type curve matching pumping test technique (Ferris et al., 1962, p. 94), consider fully penetrating pumped and observation wells without wellbore storage in a confined nonleaky aquifer infinite in areal extent. Values of dimensionless drawdown $W(u)$ are calculated and plotted on the y axis of a double-logarithmic graph against values of the dimensionless time $1/u$ on the x axis to describe a single type curve trace. Values of measured drawdown in a single observation well are plotted along the y axis on a double-logarithmic graph of the same scale as that used to plot the type curve trace against values of time on the x axis to describe a time-drawdown curve.

$W(u)$ is related to $1/u$ in the same manner that measured drawdown is related to time, thus, the measured time-drawdown curve is analogous to the type curve. The measured time-drawdown curve graph is superposed over the type curve trace keeping the $W(u)$ axis parallel to the measured drawdown axis and the $1/u$ axis parallel to the measured time axis. The measured time-drawdown curve graph is moved until it matches the type curve trace. In the matched position, a common match point for the two graphs is selected, and the four match point coordinates $W(u)$, $1/u$, s, and t are used to calculate the transmissivity and storativity of the aquifer. Transmissivity is calculated first and then storativity. This technique is essentially an exercise in visually finding the best-fit curve to scattered data.

Consider a family of type curve traces associated with fully penetrating pumped and observation wells with pumped wellbore storage in a confined leaky aquifer infinite in areal extent. The type curve trace argument is as follows (Hantush, 1960, p. 3716):

$$\tau = r/4[K'S'/(b'TS)]^{0.5} \qquad (5.1)$$

where r is the distance between the pumped and observation wells, K' is the confining unit vertical hydraulic conductivity, S' is the confining unit storativity, b' is the confining unit thickness, T is the aquifer transmissivity, and S is the aquifer storativity.

A single well measured time-drawdown graph is superposed over the family of type curves graph and moved toward and matched to a particular type curve trace. In the matched position, a common match point for the two curves is selected, and the four match point coordinates $W(u)$, $1/u$, s, and t are used to calculate the aquifer transmissivity and storativity. Transmissivity is calculated first and then storativity. The value of τ for the selected type curve trace found to be analogous to the measured time-drawdown data, previously calculated aquifer transmissivity and storativity values, and an estimated confining unit storativity are used to determine the confining unit vertical hydraulic conductivity.

Single plot type curve matching slugged well analysis proceeds as follows. A family of type curves for a range of α values is plotted on a semilogarithmic graph with values of dimensionless normalized head [$W(\alpha,\beta)$] on the arithmetic y axis and values of dimensionless time (β) on the logarithmic x axis. Values of H/H_0 are plotted on another semilogarithmic graph with the same scale on the arithmetic y axis against values of time on the logarithmic x axis to describe a time-normalized head curve. The H/H_0 and t graph is superposed over the $W(\alpha,\beta)$ and β graph. The graphs are moved horizontally along the x axis to a position where most of the data curve falls on one of the family of type curves. In the matched position, the β and t x coordinates of a common match point and the α value associated with the matched type curve are substituted in dimensionless time-normalized head equations to calculate the aquifer horizontal hydraulic conductivity and the storativity.

Conventional slugged well type curves that relate dimensionless normalized head (H/H_0) to dimensionless time (β) are very similar in shape. A determination of storativity by matching dimensionless normalized head and time test data to these type curves has a questionable reliability. However, conventional slugged well type curves that relate the first derivative of dimensionless normalized head with respect to the natural logarithm of dimensionless time differ appreciably in shape and amplitude and are strongly influenced by storativity (Karasaki et al., 1988; Spane and Wurster, 1993). Therefore, the reliability of storativity estimates can be increased when derivative type curves and test data are matched in slugged well data analysis.

Slugged well analysis with derivative type curves and test data proceeds as follows. A family of derivative type curves for a range of α values is plotted on a semilogarithmic graph with values of derivative $W(\alpha,\beta)$ on the arithmetic y axis and values of β on the logarithmic x axis. Values of derivative H/H_0 are plotted on another semilogarithmic graph with the same scale on the arithmetic y axis against values of time on the logarithmic x axis to describe a time-normalized head curve. The derivative H/H_0 and t graph is superposed over the derivative $W(\alpha,\beta)$ and β graph. The graphs are moved horizontally along the x axis to a position where most of the data curve falls on one of the family of type curves. In the matched position, the β and t x coordinates of a common match point and the α value for the matched type curve are substituted in dimensionless time-normalized head equations to calculate the aquifer horizontal hydraulic conductivity and the storativity.

SINGLE PLOT STRAIGHT-LINE MATCHING WITH INTERACTIVE CALIBRATION

The single plot straight-line matching pumping test technique (Cooper and Jacob, 1946, pp. 526–534) is based on the following confined nonleaky aquifer equation (Theis, 1935, pp. 519–524):

$$s = QW(u)/(4\pi T) \tag{5.2}$$

where

$$u = r^2S/(4Tt) \tag{5.3}$$

and s is drawdown, Q is the pumped well discharge rate, $W(u)$ is dimensionless drawdown, u is the dimensionless time, T is the aquifer transmissivity, r is the distance between the pumped and observation wells, S is the aquifer storativity, and t is the elapsed time after pumping started.

The single plot straight line technique takes advantage of the fact that semi-logarithmic graphs of arithmetic $W(u)$ vs. the logarithm of u or arithmetic $W(u)$ vs. the logarithm of $1/u$ describe a straight line when $u \leq 0.02$ and wellbore storage is negligible. u becomes ≤ 0.02 shortly after pumping starts especially at the pumped well ($r = r_w$). Arithmetic values of measured drawdown for a particular distance from the pumped well are plotted on semilogarithmic paper against the logarithms of measured elapsed time after pumping started to yield a measured time-drawdown graph. A straight line is fitted to the portion of the measured time-drawdown graph where $u \leq 0.02$ and is extended to the zero-drawdown graph axis. The slope and zero-drawdown intercept of the straight line are substituted into the following equations to calculate aquifer parameter values (Cooper and Jacob, 1946, pp. 526–534):

$$T = 2.3Q/(4\pi\Delta s) \tag{5.4}$$

$$S = 2.25T\, t_0/r^2 \tag{5.5}$$

where Q is the pumped well discharge rate, T is the aquifer transmissivity, S is the aquifer storativity, r is the distance from the pumped well to the observation well, Δs is the drawdown per logarithmic cycle (slope of the straight line), and t_0 is the zero-drawdown intercept of the straight line.

Alternatively, arithmetic values of measured drawdown at a particular measured elapsed time after pumping started are plotted on semilogarithmic paper against the logarithms of distance from the pumped well to yield a distance-drawdown graph. A straight line is fitted to the portion of the measured distance-drawdown graph where $u \leq 0.02$ and is extended to the zero-drawdown graph axis. The slope and zero-drawdown intercept of the straight line are substituted into the following equations to calculate aquifer parameter values (Cooper and Jacob, 1946, pp. 526–534):

$$T = 2.3Q/(2\pi\Delta s) \tag{5.6}$$

$$S = 2.25Tt/r_0^2 \tag{5.7}$$

where Q is the pumped well discharge rate, T is the aquifer transmissivity, S is the aquifer storativity, Δs is the drawdown per logarithmic cycle (slope of the straight line), t is the elapsed time after pumping started, and r_0 is the zero-drawdown intercept of the straight line.

After tentative values of transmissivity and storativity have been calculated, the segment of the data where $u \leq 0.02$ is determined and compared with the segment of data through which the straight line was drawn. The time t_{sL} that must elapse before the straight line technique can be properly applied to pumping test data is as follows (Walton, 1962, p. 9):

$$t_{sL} = r^2S/(0.08T) \tag{5.8}$$

where r is the distance from the pumped well to the observation well, S is the aquifer storativity, and T is the aquifer transmissivity.

The time that must elapse before a measured time-drawdown semilogarithmic graph for an observation well yields a straight line may vary from several minutes under confined nonleaky aquifer conditions to more than one day under unconfined aquifer conditions. Early measured time-drawdown data or distance-drawdown data for wells at large distances from the pumped well are filtered from the analysis when the straight line technique is used.

COMPOSITE PLOT MATCHING WITH INTERACTIVE CALIBRATION

Stallman (1971), Weeks (1977), van der Kamp (1985), and Moench (1994) discuss the importance of analyzing drawdown data for all pumping test wells on a composite plot. In the composite plot matching technique, a double-logarithmic composite plot of measured drawdown against t/r^2 is created where t is the elapsed time and r^2 is the square of the radial distance of the observation point from the axis of the pumped well (Barlow and Moench, 1999). Theoretical dimensionless time-drawdown type curves for all pumping test wells are calculated with trial aquifer parameter values. A double-logarithmic graph composite plot of the type curves is created with the same scale as the composite plot of measured drawdown vs. t/r^2.

Dimensionless drawdown h_{dd} is defined as:

$$h_{dd} = 4\pi Ts/Q \tag{5.9}$$

Dimensionless time is defined as:

$$t/r_D{}^2 = 4Tt/r^2S_{cu} \tag{5.10}$$

where

$$r_D{}^2 = r/r_w \tag{5.11}$$

and S_{cu} is the confined aquifer storativity or in the case of an unconfined aquifer with emphasis on late time-drawdown data the unconfined aquifer specific yield, T is the aquifer transmissivity, s is drawdown, Q is the pumped well discharge rate, t is elapsed time, b is the aquifer thickness, r is the distance from the pumped well to the observation well, and r_w is the pumped well effective radius.

The composite plot of measured data is superimposed onto and matched to the composite plot of type curves. The coordinate axes of each graph are kept parallel during the matching process. A composite plot match point is selected on the superimposed graphs and aquifer parameters are calculated with match point coordinates and Equation 3.262 and Equation 3.263. If the match is not satisfactory, the process is repeated with adjusted aquifer parameter values.

Alternatively, a double-logarithmic composite plot of measured time-drawdown values for all pumping test wells is constructed. Theoretical dimensional time-drawdown values for all pumping test wells are calculated with trial parameters values and plotted on the graph. The measured and calculated time-drawdown values are compared. If the comparison for all pumping test wells is not satisfactory, the conceptual model is adjusted and the process is repeated.

In the case of slug test, a semilogarithmic composite plot of measured time-normalized head values for the slugged well and nearby observation wells is constructed. Theoretical dimensional time-normalized head values for all slug test wells are calculated with trial aquifer parameters values and plotted on the graph. The measured and calculated time-normalized head values are compared. If the comparison for all slug test wells is not satisfactory, the conceptual model is adjusted and the process is repeated.

COMPOSITE PLOT AUTOMATIC PARAMETER ESTIMATION

In the composite plot with automatic parameter estimation (nonlinear regression) technique, differences between simulated drawdowns or normalized heads based on the conceptual models with trial parameter values and measured drawdowns or normalized heads are minimized using a weighted sum of squared errors objective function (Doherty, 1994; Poeter and Hill, 1997; and Hill, 1998). Aquifer test mathematical modeling equations are run repeatedly, automatically varying aquifer parameter values in a systematic manner from one run to the next until the objective function is minimized and the best fit between measured and simulated drawdowns or normalized heads is found. Statistics including sensitivities, parameter standard deviations and correlations, and drawdown or normalized head standard deviations are calculated showing the precision of calculated parameter values.

The aquifer test mathematical modeling equations to be run repeatedly must account for all the physical processes that influence measured drawdowns or normalized heads. Otherwise, the parameter estimation algorithm treats differences between measured and simulated drawdowns or normalized heads as errors in measurement with a subsequent degradation in the validity of simulated parameter values.

The person analyzing aquifer test data selects the mathematical modeling equations to be run and specifies the measured drawdowns or normalized heads, initial parameter values, and the lower and upper ranges of allowable adjustable parameter values based on conceptual models. Automatic parameter estimation results can vary depending on the group of parameters selected to be optimized and on the selected optimization time period.

Automated methods for pumping test analysis and calibration are described by Saleem (1970), Chandler et al. (1981), Grimestad (1981), Das Gupta and Joshi (1984), Mukhopadhyay (1985, 1988), Kashyap et al. (1988), Johns et al. (1992), and Cheng (2000). Kinzelbach (1986, pp. 142–151) describes a simple nonlinear regression method for automated analysis and calibration of confined nonleaky aquifer pumping test data. The computer code for the method written in Basic is provided.

Moench et al. (2001) describes the automated analysis and calibration of pumping test data by nonlinear least squares using PEST (Doherty, 1994) universal parameter estimation code. Barlow and Moench (1999) describe the automated analysis and calibration of pumping test data by nonlinear least squares using the universal UCODE (Poeter and Hill, 1998) parameter estimation code. Both PEST and UCODE require several input files and produce several output files.

WELL FUNCTION OR DRAWDOWN/HEAD CALCULATION

Well functions are classified as special functions in mathematics. Well function values are either obtained from tables in publications or are calculated with computers and polynomial approximations. The use of tables is cumbersome because interpolation is required.

Analytical integral well function values of dimensionless drawdown [$W(u)$] for the practical range of dimensionless time (u) are presented by Ferris et al. (1962, pp. 96–97) and Kruseman and de Ridder (1994). Values of dimensionless head [$W(\alpha,\beta)$] for different values of α and dimensionless time (β) are provided by Cooper et al. (1967) and Papadopulos et al. (1973). Values of dimensionless drawdown [$W(u, b)$] for the practical range of dimensionless time (u) and b are presented by Reed (1980) and Kruseman and de Ridder (1994). A computer program for calculating factorials ($m!$) is listed by Clark (1987, p. 1.15). Values of dimensionless drawdown [$W(u_a, u_b, \beta, \sigma)$] are presented by Neuman (1975a, pp. 332–333) and Kruseman and de Ridder (1994). Values of dimensionless drawdown [$W_{pp}(u\ldots)$] for selected well partial penetration conditions and assuming $s_{pp} = QW_{pp}(u\ldots)/(2\pi T)$ are presented by Weeks (1969, pp. 196–214) and Kruseman and de Ridder (1994).

Well function values are usually calculated with analytical integral mathematical modeling equations or analytical Stehfest algorithm mathematical modeling equations and a software program such as WTAQ. Pumped wellbore storage and skin, observation well delayed response, and delayed drainage at the water table effects are not usually covered in analytical integral mathematical modeling equation well functions. In addition, well partial penetration effects are either incompletely or not covered in analytical integral mathematical modeling equation well functions. However, these effects are fully covered in analytical Stehfest algorithm mathematical modeling equation well functions, which are calculated with computers and polynomial approximations. As will be explained later, well function values can also be calculated with numerical mathematical modeling equations incorporated in software programs such as MODFLOW.

CALIBRATION

There are two types of calibration: interactive and automatic parameter estimation. Interactive calibration usually precedes and guides automatic parameter estimation.

During calibration, the differences between measured and calculated drawdowns are minimized. However, calibration in itself does not ensure that the conceptual model is an accurate representation of the groundwater system and its processes (see Reilly and Harbaugh, 1993a, 1993b). Keep in mind that calibration can provide the best parameters for an ill-conceived conceptual model. The appropriateness of the conceptual model is frequently more important than achieving the smallest differences between measured and calculated drawdown. For these reasons, it is important that a careful evaluation of calibration results be made. Calibration evaluation may reveal that some calibrated parameters are not reasonable based on other knowledge of their values.

INTERACTIVE CALIBRATION

Interactive calibration rationalizes the validity of conceptual models and guides the matching of data to type curves or straight lines. Interactive calibration is a successive approximation technique involving the calculation of time-drawdown or time-normalized head values for each well based on trial conceptual models. In general, interactive calibration should evolve from simple to more complex trial conceptual models and from coarse to fine changes in trial conceptual modeling parameters thereby providing important information on the sensitivity of calculated time-drawdown or time-head values to parameter changes. During interactive calibration, calculated time-drawdown or time-head values are compared with corresponding measured time-drawdown or time-head values and residuals are calculated with the following equation:

$$r_{cal} = (s_m - s_c) \tag{5.12}$$

where r_{cal} is the residual drawdown or normalized head, s_m is the measured drawdown or normalized head, and s_c is the calculated drawdown or normalized head.

The mean absolute error of residuals, *MAE*, or the root-mean-squared error, *RMS* (or standard deviation) of residuals is calculated with the following equations (Duffield et al., 1990; Ghassemi et al., 1989; Konikow, 1978; and Anderson and Woessner, 1992):

$$MAE = 1/n \sum_{i=1}^{n} (r_{cal})_i \tag{5.13}$$

$$RMS = [1/n \sum_{i=1}^{n} (r_{cal})_i^2]^{0.5} \tag{5.14}$$

where n is the number of residuals.

The mean absolute and root-mean-squared errors prevent negative and positive residuals from canceling out.

The acceptable ranges of r_{cal}, *MAE*, and *RMS* (calibration target windows) should be predetermined and included in the conceptual model based on information concerning measurement errors and aquifer test conditions. Frequently, r_{cal} and *MAE* values within ± 5% of measured time-drawdown or time-head values are considered acceptable. Calibration is sometimes deemed acceptable when calculated time-drawdown or time-head values are within two standard deviations of calculated time-drawdown or time-head values.

Graphic measures of calibration error are helpful in determining the acceptability of calibrated parameters. Arithmetic graphs of measured drawdown or head vs. calculated drawdown or head should form a 45° line passing through the origin. Arithmetic graphs of residual vs. calculated drawdown or head should form a uniform horizontal band centered on 0. Arithmetic graphs of residual vs. time should form a uniform horizontal band centered on 0. A map of residuals in x, y, z space should exhibit a random pattern of positive and negative as well as large and small residuals.

If the interactive calibration is deemed unacceptable, the conceptual model is adjusted, the aquifer test data is reanalyzed, and the calibration process is repeated. Several conceptual model adjustments may be required before residuals and *MAE* or *RMS* are deemed acceptable. When residuals and *MAE* or *RMS* are judged to be acceptable, the conceptual model parameter values are assigned to the aquifer. The calibration process should recognize that time drawdown or time head is more sensitive to transmissivity changes than to storativity changes.

Iterative interactive calibration tasks are:

- Write input data files
- Run software program
- Read output data files
- Compare measured and output data
- Revise input data files
- Rerun software program
- Display data

Input data files can be written with word processor, spreadsheet, or aquifer test preprocessor software. Writing input data files with a word processor requires the user to become familiar with the content of WTAQ, MODFLOW, and PEST input and output data files. A preprocessor (GUI) automatically creates the input data files graphically and runs the aquifer test analysis software. The user does not need to see the input data files or know the commands that run the software until something goes wrong or the user tries to do something out of the ordinary. Then, it is important that the user understand input and output data files and run commands so that the user can track down and resolve the problem. The user must understand which data goes on each file line (record) and in which order (fields). Users can become familiar with WTAQ and MODFLOW input data files by reading instructions and examples presented by Barlow and Moench (1999) and Andersen (1993).

These documents are available at water.usgs.gov/nrp/gwsoftware/ and www.epa.gov/ada/csmos/models/modflow.html.

Regardless of whether preprocessors or word processors are used to create input data files, conceptual model and computer program solution data must be entered by the user in some manner before they are written to input data files. Data entry and file management can be quite laborious and tedious. Troubleshooting software programs can require some knowledge of the DOS language because software programs such as WTAQ, MODFLOW-96, and PEST are Fortran batch programs that run from a composite model command line containing a redirect file reference in a DOS window.

AUTOMATIC PARAMETER ESTIMATION

Automatic parameter estimation involves the preparation of PEST or UCODE input files. Detailed instructions for the preparation of PEST and UCODE input files are provided in the PEST manuals (Doherty, 2004a, 2004b) and UCODE manuals (Poeter and Hill, 1998). Briefly, three input files are required by PEST:

1. Template
2. Instruction
3. Control

Suppose analytical mathematical modeling equations incorporated in the public domain software program WTAQ are selected for analysis. In this case, the template file is simply a replica of the WTAQ input data file except the first line of the template file contains the string `ptf #` (# is a parameter delimiter) and adjustable parameters are identified in a prescribed manner. Adjustable parameters are identified by name with 1 to 12 characters. The characters cannot be the space character or the parameter delimiter character. The name is written between two parameter delimiters. There are usually 12 spaces between delimiters.

The instruction file provides instructions to PEST on how to read each calculated time-drawdown (observation) value in the WTAQ plot output plot file. The first line of the instruction file contains the string `pif @` (@ is a marker delimiter). Each succeeding line starts with a line advance item `ln` where *n* is the number of lines to advance. The line advance item is followed by a string with three white space instructions `w` separated from its neighboring instructions by one blank space. A white space instruction directs PEST to move its cursor forward from its current position until it encounters the next blank character. The white space string is followed by a string with a marker delimiter `!`, the observation name (1 to 12 characters), and a marker delimiter `!`. Thus, the instruction `l6 w w w !pwc1!` directs PEST to move its cursor down six lines then right to the third column, read the number in the third column, and store the number in the observation variable pwc1.

Many of the items in the PEST control file are used to "tune" PEST's operation to the WTAQ software. Such items include the operating mode, parameter change limits, parameter transformation types, and termination criteria. The control file contains seven sections: control data, parameter groups, parameter data, observation groups, observation data, model command line, and model input/output. The first line of the instruction file contains the string `pcf`. The second line contains the string `* control file`.

Automatic parameter estimation with numerical mathematical modeling equations incorporated in the public domain software program MODFLOW is more complex than automatic parameter estimation with analytical mathematical modeling equations incorporated in WTAQ. For example, more than one MODFLOW input file (such as Block-Centered Flow and Transient Confining Unit Leakage for confined leaky aquifer conditions) can contain parameters that require optimization. PEST template files require that each optimization parameter value be replaced with a sequence of characters (string) that identify the space as belonging to that parameter.

The uniform parameter values in input files are easily replaced with parameter space identifiers. However, nonuniform parameter values in input files can be written in the form of pumping test site grid data arrays with thousands of values and are more difficult to replace. Usually, the number of strings is limited to about 10 by zoning the pumping test site grid. A parameter value is constant within each zone.

The PEST template file can be created using the MODFLOW input file of interest and the search and replace facility of a word processor. Parameter values in a particular zone in the MODFLOW input file are altered to a parameter space identifier (such as `# hhy1 #`). This is repeated for each zone using a different parameter space identifier. Adjustable parameter arrays must be located within input files.

Instead of zoning parameter values, the PEST Pilot Points method (Doherty, 2003) can be used in automatic parameter estimation. In this method, the value of the parameter within a zone is interpolated from pilot points (scattered point set). The values of pilot points are adjusted and the surface defining the variation of hydraulic conductivity values is warped until the objective function is minimized. PEST provides an additional option for the Pilot Point method called "regularization" in which an additional measure of "stiffness" to the parameter being interpolated is imposed via a homogeneity constraint to make the inversion process more stable.

In addition, some versions of MODFLOW generate only unformatted binary output files, whereas, PEST requires an American Standard Code for Information Interchange (ASCII) model output file to operate. Thus, MODFLOW unformatted binary output files must be converted to ASCII output files before PEST can repeatedly run MODFLOW. When standard Fortran is used in the conversion process, different compilers result in unformatted files that have different structures. The files affected are those listed in a MODFLOW name file with file type DATA(BINARY) and those referenced in array-control records for the array-reading utility modules U2DINT and U2DREL where FMTIN is specified as

"(BINARY)." Thus, it is necessary that PEST and other conversion programs be compiled with the same compiler as that used to compile MODFLOW.

Further, MODFLOW generates drawdowns at specified times and at pumping test site grid block centers not necessarily at measured times and observation well sites. PEST requires that some or all of the MODFLOW generated drawdowns be interpolated to measured times and observation well sites. These tasks can be accomplished with the PEST groundwater data utility program MOD2OBS by first placing a MODFLOW command line with screen prompt redirection and then placing a MOD2OBS command line with screen prompt redirection within a batch file as part of a "composite model."

It is important that both MOD2OBS and the version of MODFLOW being used be compiled with the same compiler. MOD2OBS prompts for the names of several files, asks a series of questions, and requests data. Screen responses are redirected to a file using command line redirection so that the composite model can run continuously. The following files are required for MOD2OBS to run: settings, grid specification, bore coordinates, bore listing, measured bore sample, and calculated bore sample. The grid specification, bore coordinates, and bore listing files contain real-world space data. Measure bore sample and calculated bore sample files contain real-world time data. Conversion of measured elapsed time-drawdown data to measured real-world time-drawdown data can be time consuming. These files can be created using a word processor. The PEST instruction file is based on the MOD2OBS output file.

The MOD2OBS bore sample output file contains dates in the mm/dd/yyyy or dd/mm/yyyy format and times in the hh:mm:ss format, whereas, graphic software requires elapsed time value files. MOD2OBS bore sample output files can be converted to graphic files with the PEST groundwater data utility program SMP2HYD.

Automatic parameter estimation with MODFLOW and PEST can involve a large number (26 or more) of files. For example, suppose a pumping test automatic parameter estimation is to be performed based on a conceptual model consisting of a simple one layer aquifer with constant parameter values, one pumped well, and one observation well. Required MODFLOW oriented files would be basic (`.bas`), block-centered flow (`.bcf`), list output (`.lst`), name (`.nam`), output control (`.oc`), strongly implicit procedure (`.sip`), well (`.wel`), unformatted output (`.bin`), and MODFLOW prompt redirect (`.mrd`).

Required MOD2OBS-oriented files would be grid specification (`.gsp`), bore coordinates (`.bcd`), bore listing (`.blt`), measured bore sample (`.mbs`), calculated bore sample (`.cbs`), and MOD2OBS prompt redirect (`.ord`). Required PEST-oriented files would be settings (`.fig`), batch to run MODFLOW and PEST (`.bat`), template (`.tpl`), instruction (`.ins`), input control (`.pst`), output run record (`.rec`), output parameter values (`.par`), output parameter sensitivity (`.sen`), output observation sensitivity (`.seo`), output residuals (`.res`), and output matrix (`.mtt`).

A commercial software program Visual PEST combines PEST with the WinPEST GUI, which displays an extensive range of run-time and postrun data for easy analysis and postprocessing with PEST. Visual PEST is available at www.flowpath.com.

UCODE can also be used to analyze pumping test data with automatic parameter estimation techniques. UCODE (Poeter and Hill, 1998) was developed to:

- Manipulate model (such as WTAQ or MODFLOW) input files and read values from model output files
- Compare user-provided observations with equivalent simulated values derived from the values read from the model output files using a weighted least-squares objective function
- Use a modified Gauss–Newton method to adjust the value of user-selected input parameters in an iterative procedure to minimize the value of the weighted least-squares objective function
- Report the estimated parameter values
- Calculate and print statistics to be used to:
  - Diagnose inadequate data or identify parameters that probably cannot be estimated
  - Evaluate estimated parameter values
  - Evaluate how accurately the model represents the actual processes
  - Quantify the uncertainty of model simulated values. UCODE consists of algorithms programmed in perl, a freeware language designed for text manipulation, and Fortran90, which efficiently performs numerical calculations.

UCODE reads the following information:

- Solution control information, commands needed to execute the models, and observations from a universal file
- Instructions from a prepare file, template files, and, perhaps, a function file, which are used to create model input files with starting or updated parameter values
- Instructions from an extract file for calculating simulated equivalents for each observation from numbers extracted from the model output files

PUMPED WELL DATA

Drawdown in a pumped well can be affected by external influences as well as several or all of the following factors:

- Aquifer loss
- Well losses — linear and nonlinear
- Partial penetration loss
- Wellbore storage
- Delayed drainage at the water table under water table conditions
- Dewatering loss due to water table decline

Aquifer loss is head loss due to the laminar flow of groundwater through the aquifer toward the pumped well. This head loss is time-dependent and varies linearly with the discharge rate. Partial penetration loss is due to vertical components of flow toward a partially penetrating pumped well. This head loss is time-dependent and varies linearly with the discharge rate.

Well losses are divided into linear and nonlinear head losses. Nonlinear well losses occur inside the well screen, in the suction pipe, and in the zone adjacent to the well where the flow is turbulent. Linear well losses are caused by improvement or damage to the aquifer during drilling and completion of the well. Linear head losses are commonly referred to as *skin effect*. If the effective radius of the well is larger than the real radius of the well, the skin effect is said to be positive.

If the effective radius of the well is smaller than the real radius of the well, the skin effect is said to be negative. The skin effect is commonly defined as the difference between the total drawdown observed in the pumped well and the aquifer loss, assuming that the nonlinear well losses are negligible. The skin effect is described by the following equation (Kruseman and de Ridder, 1994, p. 216):

$$Skin\ effect = skin\ factor[Q/(2\pi T)] \quad (5.15)$$

The skin factor is defined as (Streltsova, 1988, p. 76):

$$Skin\ factor = (T/T_s - 1)\ln(r_s/r_w) \quad (5.16)$$

where Q is the pumped well discharge rate, T is the aquifer transmissivity, T_s is the near-wellbore altered transmissivity, r_w is the pumped well effective radius, and r_s is the radius of the skin near-wellbore altered transmissivity.

The following assumptions underlie procedures for analyzing pumped well data:

- Aquifer has an infinite extent
- Aquifer is homogeneous, isotropic, and uniform in thickness
- Piezometric surface is horizontal prior to pumping
- Values of u or u' are small (< 0.2)

TIME DRAWDOWN

Time-drawdown data for a pumped well can be analyzed by plotting values of adjusted drawdown s_w vs. the corresponding times from the start of the test as a semilog graph (time on logarithmic scale). A straight line is fitted through the points. The slope of the straight line (drawdown difference per log cycle of time, s_L) is determined. Aquifer transmissivity is calculated with the following equation (Cooper and Jacob, 1946):

$$T = 0.183Q/(s_L) \quad (5.17)$$

where T is the aquifer transmissivity and Q is the discharge rate.

TIME-RECOVERY

Recovery data for a pumped well can be analyzed by plotting values of residual drawdown vs. t/t as a semilog graph (t/t on a logarithmic scale). A straight line is fitted to the points. The slope of the line per log cycle s_L is determined. Aquifer transmissivity is calculated with Equation 5.17 (Kruseman and de Ridder, 1994, pp. 194–195).

SPECIFIC CAPACITY

Specific capacity data for a pumped well can be adjusted for any well losses and effects of well partial penetration. Next the aquifer transmissivity can be calculated based on the adjusted specific capacity data (Q/s_{ad}) and the following equation (Driscoll, 1986, p. 102):

$$T = F_{sc}(Q/s_{ad}) \tag{5.18}$$

where F_{sc} = 1.4 for confined nonleaky conditions, F_{sc} = 1.0 for confined leaky conditions, and F_{sc} = 0.8 for unconfined conditions. T is the aquifer transmissivity.

STEP DRAWDOWN

Step-drawdown data can be analyzed by assuming well loss to be equal to CQ^P and calculating values of H_n for each step and time of drawdown measurement with the following equation (Eden and Hazel, 1973):

$$H_n = \sum_{i=1}^{n} \Delta Q_i \log(t - t_i) \tag{5.19}$$

where n is the number of steps, ΔQ_I is the step i discharge rate increment, t is the time at which the ith step begins, and t_I is the time since the step-drawdown test started.

Values of drawdown increment s_{on1} or s_{on2} for each step are plotted vs. the corresponding calculated values of H_n as an arithmetic graph. Parallel straight lines are fitted through each set of points. The slopes of the straight lines are calculated together with the aquifer transmissivity.

The lines are extended until they intercept the $H_n = 0$ axis. The values of s_{on1} or s_{on2} at the intersections A_I are labeled. The ratio A_i/Q_i for each step (for each Q_I) is calculated. The values of A_i/Q_i are plotted vs. the corresponding values of Q_i as an arithmetic graph. A straight line is fitted through the points. The line is extended until it intercepts the ratio $A_i/Q_i = 0$ axis. The value of a is the value of A_i/Q_i at the interception. Calculated values of $(A_i/Q_i - a)$ are plotted vs. corresponding values of Q_i as a log-log graph. A straight line is fitted to the points. The slope of the straight line is determined, which is equal to $P - 1$, where P is the power to which the discharge rate must be raised to calculate well loss (P usually varies between 1.5 and 3.5 and commonly is 2). The intersection of the extended straight line with the ordinate where $Q_i = 0$ is the value of the well loss constant C.

TIME-DISCHARGE CONSTANT DRAWDOWN

Time-discharge constant drawdown data s_w for a flowing well can be analyzed by plotting values of s_w/Q vs. time t as a semilog graph (t on logarithmic scale). A straight line is fitted through the points. The straight line is extended until it intercepts the time axis where s_w/Q is 0 at the point t_0 where drawdown is zero. The slope of the straight line per log cycle of time (Δs) is determined. Aquifer transmissivity T and storativity S can be calculated with the following equations (Kruseman and de Ridder, 1994, pp. 230–231):

$$T = 2.30/(4\pi\Delta s) \tag{5.20}$$

$$S = 2.25Tt_0/r_w^2 \tag{5.21}$$

where T is the aquifer transmissivity, Q is the pumped well discharge rate, Δs is the aquifer thickness, and r_w is the well radius.

BOUNDARY DATA

The type curve matching technique can be extended to cover aquifer boundaries with the image well theory. A boundary can be diagnosed by noting any persistent time-drawdown data deviation from infinite aquifer type curve traces. Derivative time-drawdown graphs can be particularly helpful in this respect (Spane and Wurster, 1993, pp. 814–822). Barrier boundaries cause the derivative time-drawdown graph to plunge, and recharge boundaries cause the derivative time-drawdown graph to become horizontal. The type curve deviation or successive approximation method can be used to analyze boundary effects.

In the case of a barrier boundary and type curve deviation method, the appropriate infinite aquifer type curve trace is matched to early time-drawdown data unaffected by the barrier boundary and aquifer parameter values are calculated as described earlier. The infinite aquifer type curve trace is moved up and to the right and rematched to later time-drawdown data affected by the barrier boundary. The correctness of the match position is judged by noting that the drawdown value underlying a selected well function value in the early data unaffected by the barrier boundary should be half the drawdown value underlying that same selected well function value in later time-drawdown data affected by the barrier boundary.

The drawdown difference between the extrapolated first type curve trace and the second type curve trace is determined for a selected time. The aquifer parameter values calculated with early time-drawdown data unaffected by the barrier boundary, the pumped well discharge rate, the drawdown difference, and selected time values are substituted in the following equations to calculate the distance between the observation well and the image well associated with the barrier boundary:

$$W(u_i \ldots) = 4\pi T s_i / Q \tag{5.22}$$

$$r_i^2 = 4Tt_i / S \tag{5.23}$$

where $W(u_i \ldots)$ is the well function type curve trace value corresponding to the drawdown difference between the extrapolated first type curve trace and the second type curve trace s_i at a selected time t_i, r_i is the distance between the observation well and the image well associated with the barrier boundary, u_I is the value of u corresponding to $W(u_i \ldots)$, T is the aquifer transmissivity, Q is the pumped well discharge rate, and S is the aquifer storativity.

Suppose there were two barrier boundaries. The first barrier boundary is located as described earlier. The infinite aquifer type curve trace is then moved further up and to the right and matched to later time-drawdown data affected by both boundaries. The correctness of the match position is judged by noting that the drawdown value underlying a selected well function value in the early time-drawdown data unaffected by the barrier boundaries is one-third the drawdown value underlying that same selected well function value in late time-drawdown data affected by both barrier boundaries. The difference between the second and third infinite aquifer type curve traces is determined for a selected time. The aquifer parameter values calculated with early time-drawdown data unaffected by the barrier boundaries, the pumped well discharge rate, the drawdown difference, and selected time values are substituted in Equation 5.22 and Equation 5.23 to calculate the distance between the observation well and the image well associated with the second barrier boundary. The process can be repeated for other boundary image wells.

In the type curve successive approximation method, the types and locations of boundaries are estimated and corresponding trial finite aquifer type curve traces are generated using the image well theory. The time-drawdown data are matched to the trial finite aquifer type curve traces paying particular attention to the early time-drawdown data that is unlikely to be affected by boundaries. If the entire range of time-drawdown data does not match a trial finite aquifer type curve trace, the locations of the boundaries are changed and new trial finite aquifer type curve traces are generated. The process is repeated until the entire range of time-drawdown data can be matched to a trial finite aquifer type curve trace. At that point, the trial boundary locations are assigned to the aquifer.

Often, available hydrogeologic information is sufficient to determine the direction from a pumping well to a boundary, and the boundary can be located by scribing an arc with its center at the observation well and its radius equal to the distance from the observation well to the image well simulating the boundary.

A line is drawn connecting the pumping well and the arc in the appropriate direction indicated by hydrogeologic information. The image well is located at the intersection of the line and the arc. The strike of the boundary is located by bisecting the line between the pumping well and image well and drawing a line perpendicular to the line between the pumping and image wells.

If available hydrogeologic information is not sufficient to determine the direction of the boundary, at least three observation wells are required to locate the boundary. After the distances from the observation wells to the image well are calculated, arcs are scribed with their centers at the observation wells and their radii equal to the distances to the image well. The image well is located at the intersection of the arcs, and the strike of the boundary is located by bisecting the line between the pumping and image wells and drawing a line perpendicular to the line between the pumping and image wells.

MODFLOW SIMULATION

A full description of MODFLOW groundwater flow simulation is presented by McDonald and Harbaugh (1988). In addition, general groundwater flow modeling details are provided by Anderson and Woessner (1992). What is not covered in these two documents and is presented herein are details concerning the simulation of pumping tests with MODFLOW.

Pumping test data analysis with MODFLOW, similar to pumping test data analysis with analytical models, involves the selection of an appropriate analysis format and technique. Usually, the dimensional format is selected with interactive and automated comparison of calculated and measured time-drawdown or head values (model calibration) and the minimization of differences between calculated and measured time-drawdowns or heads (residuals) interactively and automatically with nonlinear regression. First, calculated and measured time-drawdowns for each single well are compared using the interactive calibration technique and then calculated and measured time-drawdowns or heads for all wells are compared using the automatic parameter estimation technique.

The dimensionless format involves the generation of several dimensionless time-drawdown type curve traces covering the range of possible pumping test site conceptual models with MODFLOW. Measured time-drawdowns values are matched to the family of type curves as described for analytical models.

ABOUT MODFLOW VERSIONS

There are four major versions of MODFLOW that can be used to simulate pumping tests: MODFLOW-88, MODFLOW-96, MODFLOWP-96, and MODFLOW-2000. The first two versions solve only groundwater flow equations, the third version solves groundwater flow and parameter estimation equations, and the fourth and latest version solves groundwater flow and other equations such as parameter estimation and solute transport. MODFLOW-2000 and MOD-

FLOWP-96 internally support both aquifer test interactive calibration and automatic parameter estimation. MODFLOW-88 or MODFLOW-96 internally support aquifer test interactive calibration and when supplemented externally with UCODE or PEST support automatic parameter estimation. MODFLOW-88 and MODFLOW-96 support transient leakage in confining units (TLK1 package), whereas, MODFLOWP-96 and MODFLOW-2000 do not. Both areal and cylindrical flow to a well during a pumping test can be simulated with MODFLOW-88 and MODFLOW-96. The short duration (several minutes or less) and radius of influence (tens of feet) of a slug test make it impractical to simulate slug test data with MODFLOW.

MODFLOW-88 and MODFLOW-96 like MODFLOW-2000 and MODFLOWP-96 can be used to analyze pumping test data with interactive calibration. In addition, time-drawdown values calculated with MODFLOW-2000, MODFLOW-96, or MODFLOW-88 can be used with automatic parameter estimation software such as PEST to analyze pumping test data.

Without the Parameter-Estimation package, MODFLOWP-96 can be used to interactively analyze pumping test data. With the Parameter-Estimation package, MODFLOWP-96 can be used to estimate parameters by nonlinear regression. Parameters used to compute the following MODFLOW model inputs can be estimated:

- Layer transmissivity, storage, coefficient of storage, hydraulic conductivity, and specific yield; vertical leakance; horizontal and vertical anisotropy
- Hydraulic conductance of the River, Streamflow-Routing, General-Head Boundary, and Drain packages
- Areal recharge
- Maximum evapotranspiration
- Pumpage
- Hydraulic head at constant-head boundaries

Nearly any spatial variation in parameters can be defined by the user. Data used to estimate parameters can include existing independent estimates of parameter values, observed hydraulic heads or temporal changes in hydraulic heads, and observed gains and losses along head-dependent boundaries (such as streams). Model output includes statistics for analyzing the parameter estimates and the model; these statistics can be used to quantify the reliability of the resulting model, to suggest changes in model construction, and to compare results of models constructed in different ways. Parameters are estimated by minimizing a weighted least-squares objective function by the modified Gauss–Newton method or by a conjugate-direction method.

All of the following MODFLOW-2000 are required for pumping test analysis:

- GWF1 — Ground-Water Flow process
- SEN1 — Sensitivity process

- OBS1 — Observation process
- PES1 — Parameter-Estimation process

The Ground-Water Flow process generates model-calculated drawdown values at model grid nodes. Several of the following MODFLOW-2000 Ground-Water Flow process packages files can be required for pumping test analysis:

- BAS6 — Basic
- BCF6 — Block-Centered Flow
- LPF1 — Layer-Property Flow
- RIV6 — River
- DRN6 — Drain
- WEL6 — Well
- GHB6 — General Head Boundary
- RCH6 — Recharge
- EVT6 — Evapotranspiration
- CHD6 — Time-Variant Specified Head
- SIP5 — Strongly Implicit Procedure
- SOR5 — Slice Successive Over-Relaxation
- PCG2 — Version 2 of Preconditioned Conjugate Gradient
- DE45 — Direct solver
- LMG1 — Multigrid solver
- FHB1 — Flow and Head Boundary
- HUF1 — Hydrogeologic-Unit Flow (with horizontal anisotropy capability)
- LAK3 — Lake
- DRT1 — Drains with Return Flow

The Observation process generates model-calculated drawdown values at observation well sites for comparison with measured drawdown values. A variety of statistics is calculated to quantify this comparison, including a weighted least-squares objective function. In addition, a number of files are produced that can be used to compare the values graphically. The Sensitivity process calculates the sensitivity of hydraulic heads throughout the model with respect to specified parameters using the accurate sensitivity-equation method. These are called grid sensitivities. If the Observation process is active, it uses the grid sensitivities to calculate sensitivities for the simulated values associated with the observations. These are called observation sensitivities. Observation sensitivities are used to calculate a number of statistics that can be used to diagnose inadequate data, to identify parameters that probably cannot be estimated by regression using the available observations, and to evaluate the utility of proposed new data.

The Parameter-Estimation process uses a modified Gauss–Newton method to adjust values of user-selected input parameters in an iterative procedure to minimize the value of the weighted least-squares objective function. Statistics produced by the Parameter-Estimation process can be used to evaluate estimated parameter values; statistics produced by the Observation process and postprocessing program RESAN-2000 can be used to evaluate how accurately the model represents the actual processes; statistics produced by postprocessing program YCINT-2000 can be used to quantify the uncertainty of model simulated values. MODFLOW-2000 has no regularization mode and does not in itself support parametrization through the use of pilot points.

A special version of MODFLOW-2000, modified for optimal use of PEST, and known as MODFLOW-ASP and a MODFLOW2000-to-PEST translator are available at www.sspa.com. Using the translator, an input data set for MODFLOW-2000's parameter estimation process can be converted to a MODFLOW–PEST input data set thereby making it possible to use PEST's advanced regularization and pilot points functionality with MODFLOW-2000.

Parameters are defined in the Ground-Water Flow process input files and can be used to calculate most model inputs, such as for explicitly defined model layers, horizontal hydraulic conductivity, horizontal anisotropy, vertical hydraulic conductivity or vertical anisotropy, specific storage, and specific yield; and, for implicitly represented layers, vertical hydraulic conductivity.

In addition, parameters can be defined to calculate the hydraulic conductance of the River, General-Head Boundary, and Drain packages; areal recharge rates of the Recharge package; maximum evapotranspiration of the Evapotranspiration package; pumpage or the rate of flow at defined-flux boundaries of the Well package; and the hydraulic head at constant-head boundaries.

MODFLOW-2000 reads input from the following files:

- Ground-Water Flow process package input files, which define the groundwater flow simulation and parameters that can be listed in the Sensitivity process input file
- Observation process input files, which define the observations
- Sensitivity process input file, which lists the parameters for which
    - Values are controlled by the Sensitivity process
    - Sensitivities are to be calculated
    - Values are to be estimated through the Parameter-Estimation process
- Parameter-Estimation process input file, which lists values for variables that control the modified Gauss–Newton nonlinear regression

Detailed instructions for the preparation of process input files are provided in MODFLOW documentation. Input files can be created with a word processor or one of the many available public domain or commercial preprocessors. For example, MFI2K (Harbaugh, 2002) is a data-input (entry) Fortran program for MODFLOW-2000 developed by the U.S. Geological Survey. MFI2K is designed to be easy to use; data are entered interactively through a series of display screens. MFI2K also can be used in conjunction with other data-input programs so that the different parts of a model data set can be entered using the most suitable program. MFI2K interfaces to an external program such as Microsoft® Excel® for entering or editing two-dimensional arrays and lists of stress data.

HYDMOD, a utility program developed by the U.S. Geological Survey (Hanson and Leake, 1999) saves time-drawdown series information at user-specific locations from a MODFLOW simulation thus assisting in interactive calibration. HYDMOD does not calculate time-drawdown series information at user-specific measurement times. HYDMOD uses the additional utility subroutine URWORD that is included in MODFLOW-96 and MODFLOW-2000 (Harbaugh and McDonald, 1996). This additional subroutine must be added to earlier versions of MODFLOW to use HYDMOD. Four steps are required to create time-drawdown series data from MODFLOW using HYDMOD. First, the user creates an input data set that specifies the locations and types of output data desired from a simulation. Second, MODFLOW is run with HYDMOD to create a binary file of the specified data at the specified locations. Third, the binary file is processed with the postprocessor program HYDFMT into an ASCII-text data file. And, fourth, the data can be plotted directly or entered into a spreadsheet. As a quick alternative to this approach, HYDPOST (a MODFLOW postprocessor program) can extract time-drawdown series data from the unformatted (binary) output files generated by MODFLOW without the use of HYDMOD.

The MODFLOW-96 word Output Control file controls the amount, type, and frequency of information to be printed or written to formatted or unformatted (binary) files. The default output consists of head values and budgets terms printed for the end of each stress period. Additionally, if starting heads are saved, drawdown is printed at the end of every stress period. Printed output actually means MODFLOW-96 output in ASCII-text readable form that might possibly be sent to a printer. Such output including model data and calculated heads, drawdowns, and cell-by-cell flow terms is actually written to a file called the Listing file for use prior to the actual printing. The Listing file can be edited and printed with a commercial word processor such as Microsoft Word.

MODFLOW-96 formatted data files contain formatted arrays records dimensioned (NCOL, NROW). These records can be preceded by identifying informational records if so specified in the Output Control file. A formatted data file containing z drawdown values for a single stress period and time step can be generated by specifying appropriate data in the Word Output Control file and used to create a contour map or a cylindrical well cross section.

MODFLOW-96 formatted output data for a specified time is not in a form suitable for creating contour maps. Map graphics programs require XYZ data. MODFLOW-96 cell center real-world coordinates can be calculated with the PEST utility program GRID2PT and then added to the MODFLOW-96 output grid z data to generate graphics XYZ data.

MODFLOW-2000 supports the use of nonstandard options in `OPEN` statements for unformatted files. This is useful on personal computers because it makes it possible to use different compilers and still have MODFLOW produce unformatted files with the same structure. This causes the structure of unformatted files to be different than what was used in previous versions of MODFLOW. The new files are referred to as "unstructured" because they eliminate the vendor-specific structure normally included in such files. Programs such as PEST that read unformatted files produced by MODFLOW-2000 must be able to read unstructured, unformatted files (using `OPEN`-statement specifiers and options appropriate for the compiler being used).

PUMPED WELL DRAWDOWN

A pumped well is simulated in MODFLOW by imposing a discharge rate on a grid block node. The grid block node is usually much larger than the pumped well diameter. The drawdown calculated by MODFLOW at the pumped well node is an average drawdown for the grid block, not the drawdown in the pumped well (Beljin, 1987, pp. 340–351). The concept of equivalent well block radius with a square (uniform) or rectangular (variable) grid and anisotropic aquifer hydraulic conductivity (Peaceman, 1983, pp. 531–543) is used to estimate the drawdown in the pumped well based on the drawdown calculated by MODFLOW.

The radius at which the drawdown in the aquifer is equal to the drawdown calculated by MODFLOW for the pumped well grid block is estimated with the following equation (Peaceman, 1983):

$$r_e = 0.28[(T_{yy}/T_{xx})^{0.5}\Delta x^2 + (T_{xx}/T_{yy})^{0.5}\Delta y^2]^{0.5}/[(T_{yy}/T_{xx})^{0.25} + (T_{xx}/T_{yy})^{0.25}] \quad (5.24)$$

where r_e is the equivalent well block radius, T_{xx} is the aquifer transmissivity in the x direction at the pumped well node, T_{yy} is the aquifer transmissivity in the y direction at the pumped well node, Δx is the pumped well block grid spacing in the x direction, and Δy is the pumped well block grid spacing in the y direction.

The drawdown in the pumped well based on the drawdown calculated by MODFLOW for the pumped well block node is estimated with the following equation (Peaceman, 1983):

$$s_p = s_b + Q\ln(r_e/r_w)/[2\pi(T_{xx}T_{yy})^{0.5}] \quad (5.25)$$

where s_p is the drawdown in the pumped well, s_b is the drawdown calculated by MODFLOW for the block node, r_e is the equivalent well block radius, r_w is the pumped well effective radius, and Q is the pumped well discharge rate.

A method developed by Pedrosa and Aziz (1986, pp. 611–621) may be used to estimate the drawdown in the pumped well with greater precision than can be attained with the above equations.

MULTILAYER PUMPED WELL

A multilayer (node) pumped well can be approximately simulated in MODFLOW by placing more than one well per nodal block. The total pumping rate is apportioned among these wells using the following equation (McDonald and Harbaugh, 1988):

$$Q_{i,j,k} = T_{i,j,k}(Q_T/T_{sum}) \quad (5.26)$$

where $Q_{i,j,k}$ is the pumping rate from individual layers, $T_{i,j,k}$ is the transmissivity of each layer, Q_T is the total pumping rate, and T_{sum} is the sum of the transmissivities of all layers penetrated by the well.

The effective well block radius for each layer and the effective water levels at the pumped well blocks in each layer can also be calculated with equations presented by Kontis and Mandle (1988).

A multilayered pumped well can be more accurately simulated with the Multi-Node Well (MNW) package for MODFLOW (Halford and Hanson, 2002) available at water.usgs.gov/pubs.

WELLBORE STORAGE

Pumped wellbore storage is not simulated in MODFLOW because most models do not need to simulate this short-term well feature. However, pumping test models must simulate pumped wellbore storage so that very early time-drawdown data can be accurately generated. Pumped wellbore storage with areal flow to a well can be simulated by estimating aquifer flow rates and setting MODFLOW discharge rates equal to aquifer flow rates.

During very early portions of a pumping test, part of the water withdrawn from a well is derived from water stored in the well casing (wellbore storage) and is not withdrawn from the aquifer system (aquifer flow rate). The aquifer flow rate and the associated time rate of drawdown are less at very early times than they are at later times because of wellbore storage effects. If the pumped well has a finite diameter and wellbore storage is appreciable, the discharge rate is the sum of the aquifer flow rate and the rate of wellbore storage depletion. The aquifer flow rate increases exponentially with time toward the discharge rate and the wellbore storage depletion rate decreases in a like manner to 0 (Streltsova, 1988, pp. 49–55).

Aquifer flow rates can be estimated by analytically calculating drawdowns at the pumped well for selected elapsed times with and without pumped wellbore storage using average aquifer parameter values for the pumping test site and a constant discharge rate. The constant discharge rate is multiplied by ratios of drawdowns with and without pumped wellbore storage to estimate aquifer flow rates. Drawdowns with and without pumped wellbore storage can be calculated with WTAQ.

Pumped well wellbore storage with cylindrical flow to a well can be simulated by assigning a very large (effectively infinite) radial conductance between nodes that represent the well at the well screen, zero radial conductance between nodes representing the well where the well is cased, a large (effectively infinite) vertical conductance inside the pumped well, and a storage capacity for the topmost node in the well (representing the free surface) that corresponds to a unit value of specific yield.

Observation well delayed response also is not simulated in MODFLOW because most models do not need to simulate this well feature. However, some pumping test models must simulate observation well delayed response so that early time-drawdown data can be calculated. The effects of observation well delayed response can be simulated by analytically calculating drawdowns at an observation point for selected elapsed times with and without observation well delayed response using average aquifer parameter values for the pumping test site and a constant discharge rate. Differences in drawdowns with and without observation well delayed response are subtracted from drawdowns previously calculated with MODFLOW and variable discharge rates. Drawdowns with and without observation well delayed response can be calculated with WTAQ.

DELAYED GRAVITY DRAINAGE

In unconfined aquifers, time-drawdown data usually show a typical *S*-shape composed of three distinct segments: a steep early time segment, a flat intermediate time segment, and a relatively steep late time segment (Neuman, 1975a). The first segment covers a brief period often only a few minutes in length during which the unconfined aquifer reacts in the same way as a confined aquifer. The water discharged from the well is derived from aquifer storage by the expansion of the water and the compaction of the aquifer. The second segment, which can range in length from several minutes to days, mainly reflects the impact of delayed gravity drainage of the interstices (leakage) within the cone of depression created during the first segment. The third segment reflects a period during which the water discharged from the well is derived both from gravity drainage of interstices and the expansion of the water and the compaction of the aquifer.

Delayed gravity drainage is usually not simulated in MODFLOW because most models do not need to simulate this condition. However, pumping test models must simulate gravity drainage so that very early time-drawdown data can be accurately calculated. The simulation can be accomplished by using fine discretization in time and space in an areal or cylindrical well simulation (Reilly and Harbaugh, 1993a, pp. 489–494). The unconfined aquifer is subdivided into several (usually 10 or more) layers, especially in the upper parts of the aquifer within which gravity drainage of interstices occurs depending upon the desired simulation accuracy.

Confined primary storage coefficients are assigned to all layers except the uppermost layers. Aquifer specific yields are assigned to the secondary storage coefficients for the uppermost layers. The optimal discretization of layers can be determined by varying the number of layers, simulating time-drawdown in aquifer systems with uniform parameter values, and comparing MODFLOW results and time-drawdown values calculated with analytical Stehfest algorithm modeling equations. Delayed drainage at the water table described by Moench et al. (2001) is not simulated in MODFLOW.

GRID DESIGN

A fundamental aspect of MODFLOW is the representation of the aquifer test domain by discrete volumes of material. The accuracy of the model is limited by the size of the discrete volumes. If the cell size is too large, important aquifer system features may be left out or poorly represented. Thus, it is important to incorporate a sufficient number of cells to allow the complexity of drawdown or head distribution to be simulated.

The MODFLOW grid for areal pumping test simulations typically consists of 50 rows and 50 columns within the pumping test site. The grid is aligned appropriately with respect to the pumping test site and any aquifer boundaries or discontinuities. The pumped well is usually located in the center block of the grid. Typically, square blocks of the grid have spacings of 20 to 100 ft within a radius of 1000 ft of the pumped well. The grid spacing is expanded beyond 1000 ft out to the boundaries by increasing nodal spacing no more than 1.5 times the previous nodal spacing. The grid is designed so there is negligible drawdown at the grid borders. The grid can be designed so that grid nodes are at the location of observation wells to avoid spatial interpolation of MODFLOW output during calibration.

MODFLOW output data and grid *xy* coordinates are commonly used to create a drawdown (z) contour map for a specified elapsed pumping test time. Contour map software interpolates regular grid drawdown (z) data based on irregular grid MODFLOW drawdown (z) data. Different interpolators provide different interpretations of the irregular MODFLOW drawdown (z) data. Best results are achieved where the MODFLOW grid is regular. For this reason, the MODFLOW grid should be regular near the pumped well in the vicinity of observation wells. Calibration with contour maps should proceed with due caution because all interpolators distort data to some extent.

The MODFLOW grid for cylindrical well simulations typically consists of several rows and 40 columns radially spaced with a multiplier of 1.5. The grid starts at the top row (layer thickness) and ends at the bottom row. Layer thickness for the top and bottom rows is one-half the adjoining row thickness. The pumped well is located in Row 1 and there is negligible drawdown along Column 40, which is a constant-head boundary. Under unconfined conditions, there are usually 11 to 21 rows in order to simulate slow gravity drainage under unconfined aquifer conditions.

TIME DESIGN

For transient models, time is represented by discrete increments of time called stress periods and time steps in MODFLOW. The size of the time steps has an impact on the accuracy of a model. Thus, it is important to subdivide time into a sufficient number of stress periods and time steps to allow the complexity of drawdown or head distribution to be simulated.

MODFLOW simulation time is divided into stress periods (time intervals during which the pumping rate is constant) that are, in turn, divided into time steps. The lengths of each stress period, the number of time steps into which each stress period is divided, and the ratio of the length of each time step to that of the preceding time step (time step multiplier) need to be specified.

Several stress periods each with a single time step and a 1.0 time step multiplier are typically specified for pumping test analysis. For simulation of confined aquifer conditions, time is typically subdivided into 36 stress periods, each with a specified aquifer flow rate and logarithmic spaced lengths ranging from 0.0002 to 0.10 day. For simulation of unconfined aquifer conditions, time is typically more finely subdivided than for confined aquifer conditions into 45 stress periods, each with a different aquifer flow rate and logarithmic spaced lengths ranging from .00002 to 0.1 day. Optimum logarithmic stress period lengths can best be determined with the aid of a semilogarithmic graph. A typical set of simulations times (day) are:

1×10^6	1×10^5	1×10^4	1×10^3	1×10^2	1×10^1
1.5×10^6	1.5×10^5	1.5×10^4	1.5×10^3	1.5×10^2	1.5×10^1
2×10^6	2×10^5	2×10^4	2×10^3	2×10^3	2×10^1
2.5×10^6	2.5×10^5	2.5×10^4	2.5×10^3	2.5×10^2	2.5×10^1
3×10^6	3×10^5	3×10^4	3×10^3	3×10^2	3×10^1
4×10^6	4×10^5	4×10^4	4×10^3	4×10^2	4×10^1
5×10^6	5×10^5	5×10^4	5×10^3	5×10^2	5×10^1
6×10^6	6×10^5	6×10^4	6×10^3	6×10^2	6×10^1
8×10^6	8×10^5	8×10^4	8×10^3	8×10^2	8×10^1

The accuracy of the simulation for the first few time steps may not be high (de Marsily, 1986, p. 400). The stress periods can be designed so that total elapsed times coincide with measured observation times to avoid temporal interpolation of MODFLOW output during calibration.

TRANSIENT CONFINING UNIT LEAKAGE

Confining units are not explicitly represented in MODFLOW. Instead, a standard part of MODFLOW (Block-Centered Flow package) indirectly simulates steady-state confining unit leakage without confining unit storage changes by means of a vertical leakance (vertical confining unit hydraulic conductivity divided by confining unit thickness) term known as VCONT at each finite-difference grid node. The source of water to the confined leaky aquifer may be another confined aquifer or an unconfined aquifer. However, it assumed that the head in the source unit is constant, there is no release of water from storage within the confining unit, and flow in confining units is vertical (horizontal flow in confining units is negligible). Thus, confining unit storativity is not simulated in MODFLOW because most models do not need to simulate this condition. However, pumping test models must simulate confining unit storativity so that very early time-drawdown data can be accurately calculated (see Hantush, 1964).

Transient confining unit leakage can be simulated in MODFLOW using several layers (usually less than 20) with assigned values of specific storage and leakance to represent the confining unit (McDonald and Harbaugh, 1988). Hydrologically equivalent confining unit specific storage and leakance values are assigned to individual layers in proportion to model dimensions in the simulation of transient confining unit leakage with layers. Very small transmissivities are assigned to the layers so that the flow in the confining unit is essentially vertical. The number of layers is determined by the method of successive approximations wherein the results of simulations with a given number of layers are compared with analytical solutions.

Use of one layer to simulate transient confining unit leakage results in the vertical distribution of head in the confining unit being approximated with single head value at the center of the confining unit. Addition of layers to represent the confining unit adds detail to the approximation of head value at the center of the confining unit. A source unit with a constant head (such as a surface water body) is simulated by assigning a large (0.2 to 0.5) specific yield to each source unit finite-difference grid node.

The TLK1 (Leake et al., 1994) package allows simulation of transient leakage without the use of additional model layers to simulate a confining unit. A confining unit must be bounded above and below by model layers in which head is calculated or specified. For a confining unit that pinches out, transient equations are used where the confining unit exists and VCONT terms are used where the confining unit is absent. Specific storage is assumed to be constant. The VCONT terms for layers surrounding a confining unit are set to 0.0 in the BCF package. When a transient leakage parameter at a node is set to zero or less, TLK1 does not carry out transient leakage calculations at that node. Instead, leakage is calculated using the VCONT value for that node in the BCF package. TLK1 cannot simulate transient leakage in a confining unit that is bounded on the top or bottom by an impermeable boundary nor a situation where the water table is within the confining unit. The wetting capability should not be used for any model layers that connect to a confining unit that is being simulated with the TLK package.

SPATIAL PARAMETER DEFINITION

Some horizontal anisotropic conditions can be simulated by MODFLOW by specifying appropriate anisotropy factors (TRPY) in the BCF package. TRPY is a one-dimensional array containing an anisotropy factor for each layer in a model. It is the ratio of transmissivity or hydraulic conductivity along a grid column to transmissivity or hydraulic conductivity along a grid row. TRPY is set to 1.0 for isotropic conditions.

Conceptual model spatial parameter (heterogeneity) can be defined during interactive or automatic parameter estimation MODFLOW pumping test model calibration. In interactive calibration, the MODFLOW grid within the pumping test domain can be subdivided into several zones with assigned uniform grid cell parameter values based on available data. Alternatively, parameter values can be assigned to scattered data points. Parameter values at each MODFLOW grid cell can then be interpolated from scattered data point parameter values using the Kriging method (see de Marsily, 1986) and stored in a MODFLOW input data array file.

Kriging (see Isaaks and Srivastava, 1989) is a statistical interpolation method that calculates the best linear unbiased estimate for the parameter of interest. The parameter is assumed to be a random function whose spatial correlation (structure) is defined by a variogram that relates changes in parameter value with changes in distance. Kriging provides an estimate of the interpolation error. A number of Kriging software packages based on the geostatistical software library (GSLIB) (Deutsch and Journel, 1997), including GEO-EAS and GEOPACK are available at www.scisoftware.com.

MODFLOW is run with the zoned or Kriged input data array file and calculated drawdown values are compared with measured drawdown data. If the comparison is acceptable based on calibration statistics, the spatial parameter definition as described in the input data array file is deemed optimal. Otherwise, the procedure is repeated with an adjusted number of zones or assigned parameter values until an acceptable comparison of measured drawdown data and calculated drawdown values is obtained. This procedure requires a great deal of experience and can be quite laborious and slow. The nonuniqueness of the results can be high.

Alternatively, the procedure described above can be terminated after a few exploratory iterations short of optimization but can be adequate enough to refine conceptual model parameter value initial, lower, and upper range limits. Parameter values can then be optimized automatically in two steps with PEST.

During the first step, PEST can run in either the parameter or regularization mode and adjustable parameters must be stored internally within the MODFLOW input data files. The MODFLOW grid within the pumping test domain can be subdivided into several zones with estimated uniform grid cell parameter values based on interactive calibration results. If PEST runs in the estimation mode, the number of zones must not exceed the number of observations plus the number of prior information statements. If PEST runs in the regularization mode, the number of zones can exceed the number of observations plus the number of prior information statements.

Differences between calculated and measured drawdowns are minimized using a weighted sum of squared error objectives. MODFLOW is run repeatedly while PEST automatically varies the zone adjustable parameter values in a systematic manner from one run to the next until the objective function is minimized. Statistics are provided showing the precision of calculated parameter values.

During the second step, the regularization and pilot points capabilities of PEST (Doherty, 2003) can be used to define small-scale spatial parameter value variations especially important in contaminant transport studies. The PEST regularization capability makes a pumping test domain only as heterogeneous as it needs to be to achieve an acceptable level of fit between MODFLOW calculated drawdowns and corresponding measured drawdowns. The pumping test domain need not be subdivided into a small number of zones of piecewise parameter constancy. Instead, a large number of pilot points with adjustable or fixed parameter values are scattered liberally throughout the pumping test domain with increased density in suspected heterogeneous areas, in the immediate vicinity of boundaries, and in areas where there is greater density of measurement points.

PEST is run in the regularization mode and adjustable parameters are stored external to the MODFLOW input data files.

Parameter values are assigned by PEST to each pilot point during each iteration, based in part on pilot point initial, lower, and upper parameter values specified in the PEST Control file. The pilot point parameter values are spatially interpolated to all MODFLOW grid cells within the pumping test domain using Kriging and the PEST utilities PP2FAC and FAC2REAL and pilot point, grid specification, structure, and MODFLOW-compatible integer array files. In assigning parameter values to pilot points, PEST effectively assigns parameter values to the entire MODFLOW grid. Hence, PEST is able to define heterogeneity by itself through the assignment of pilot point parameter values during the optimization process.

In the spatial interpolation, the same geostatistical structure can be assigned to all active MODFLOW cells or different geostatistical structures can be assigned to particular zones within the pumping test domain. Different geostatistical zones are identified in the MODFLOW Basic package boundary integer array. If appropriate, different subsets of pilot points can be used as a basis for spatial interpolation within different mapped geological units. For each subset of pilot points, a series of differences can be formed between the parameter values assigned to these points in order to create a regularization scheme.

PPK2FAC generates a set of Kriging factors for use in spatial interpolation from a set of pilot points to a MODFLOW finite-difference grid. Kriging factors are based on user-supplied, nested variograms, each with a specified magnitude and direction of anisotropy. Different variograms can be used for spatial interpolation in different parts of the pumping test domain. PPK2FAC also writes a MODFLOW-compatible real array depicting Kriging standard deviations over the pumping test domain, as well as a regularization information file, which can be used to introduce geostatistically based regularization constraints to calibration. FAC2REAL generates a MODFLOW parameter input array file based on PPK2FAC-generated Kriging factors. FAC2REAL writes a MODFLOW-compatible real array file.

A pilot point file contains pilot point data. The first entry on each line of a pilot point file is the pilot point identifier. The second and third entries on each line contain the easting and northing coordinates of the pilot point. The fourth entry on each line is an integer; this normally identifies the zone within a pumping test domain in which spatial interpolation is affected by the value assigned to the pilot point. The fifth entry on each line is the value assigned to the pilot point. A grid specification file contains all of the information required to define the finite-difference grid on a map using real-world coordinates. Information contained in this file is used to determine the positions of pilot points with respect to the grid and to superimpose MODFLOW outputs on other geographical data. A structure file is read by PPK2FAC to ascertain the geostatistical characteristics of the areas in which spatial interpolation is to be carried out. An integer array file holds a MODFLOW-compatible integer array.

Pilot points and regularization can be used in conjunction with stochastic fields to explore the predictive uncertainty of a calibrated pumping test conceptual model (Doherty, 2003). The stochastic field generator PEST utility program FIELDGEN in conjunction with PEST's regularization functionality can be used to generate many different parameter fields, which also calibrate a pumping test conceptual model. Predictions can be made using these parameter fields as a means of exploring the uncertainty accompanying pumping test conceptual model predictions.

Several commercial MODFLOW software packages supporting automatic parameter calibration with the regularization and pilot point capabilities of PEST are available at www.scisoftware.com.

STREAMBED-INDUCED INFILTRATION

Simulation of streambed-induced infiltration requires knowledge of the streambed vertical hydraulic conductivity as well as aquifer parameters. Because the streambed thickness is rarely known with any degree of accuracy, the hydraulic characteristic of the streambed is usually expressed as the ratio of the streambed vertical hydraulic conductivity divided by the streambed thickness and called the streambed leakance. Typical areal investigation streambed leakances known to the author are presented in Table 5.1. Streambed hydraulic conductivity data, which can be used to estimate representative upper and lower conceptual model ranges, are presented by Calver (2001). Usually, streambed hydraulic conductivities based on areal investigations range from 0.028 to 283 ft/day; whereas, streambed hydraulic conductivities based on point investigations range from 0.00028 to 2335 ft/day. Streambed thicknesses commonly range from a fraction of a foot to several feet.

Streambed leakance can be estimated with point and areal investigations. Shallow depth and small volume point streambed hydraulic conductivity investigations in stream channels are discussed by de Lima (1991); Duwelius (1996);

TABLE 5.1
Typical Streambed Leakance Values

Location	Leakance (1/day)	Water Temperature (°F)
Satsop River, Washington	30.7	51
Mad River, Ohio	3.1	39
Sandy Creek, Ohio	2.2	82
Mississippi River, Missouri	1.0	54
White River, Indiana	0.7	69
Miami River, Ohio	0.5	35
Mississippi River, Missouri	0.3	34
White River, Indiana	0.1	38
Mississippi River, Missouri	0.1	83

Chen (2000); and Landon et al. (2001). Greater depth and larger volume pumping test investigations of streambed hydraulic conductivity with the River package of MODFLOW and nonlinear regression inverse methods are described by Yager (1993) and Chen and Chen (2003).

The results of streambed-induced infiltration sensitivity analysis by Chen and Chen (2003) indicate that:

- Sensitivities for streambed hydraulic conductivity are low when pumping test times are < 0.1 day
- Sensitivities for streambed hydraulic conductivity and aquifer hydraulic conductivity increase with pumping test time
- Sensitivities for aquifer anisotropy are high when pumping test times are < 0.1 day
- Sensitivities for aquifer storativity are low when pumping test times are > 0.01 day
- Sensitivities for aquifer specific yields increase with pumping test times up to about 0.7 day when they decrease
- Sensitivities for aquifer hydraulic conductivity, anisotropy, and specific yield decrease near the stream
- Sensitivities for streambed hydraulic conductivity increase with greater pumping rates

In MODFLOW, a streambed is divided into reaches and assigned to grid cells. Streambed-aquifer interconnection is represented as a conductance block through which one-dimensional flow occurs. The conductance block has a width equal to the streambed width, a length equal to the reach length (grid spacing), and a thickness equal to the thickness of the streambed. It is assumed that there is no significant head loss between the bottom of the streambed layer and the point represented by the underlying grid cell node; the underlying grid cell remains fully saturated; the streambed storativity is negligible; and when the head in the aquifer drops below the streambed, induced streambed infiltration is no longer proportional to the aquifer head but instead is dependent on the stream stage and the streambed thickness.

The rate of induced streambed infiltration to the aquifer Q_{riv} for a unit length of the reach is (McDonald and Harbaugh, 1988):

$$Q_{riv} = C_{riv}(h_{riv} - h_{aq}) \quad (5.27)$$

with

$$C_{riv} = K_{riv}W_{riv}/M_{riv} \quad (5.28)$$

where C_{riv} is the streambed conductance, h_{riv} is the head in the stream, h_{aq} is the head in the aquifer at the grid node cell underlying the stream reach, K_{riv} is the vertical hydraulic conductivity of the streambed, W_{riv} is the width of the stream channel in the grid cell, M_{riv} is the thickness of the streambed, and K_{riv}/M_{riv} is the streambed leakance.

Streambed conductance is usually determined by assigning an initial conceptual model streambed conductance and then adjusting the initial streambed conductance during interactive or automatic calibration within conceptual model upper and lower streambed conductance limits. Data regarding the streambed is specified in the MODFLOW River package input data file. One record for each river reach and each stress period is required. The record consists of the layer number, row number, and column number of the grid cell containing the streambed reach; the head in the stream at the grid cell; the streambed conductance at the grid cell; and the elevation of the bottom of the streambed at the grid cell. Thus, it is possible to vary the streambed conductance from one grid cell to another and from one stress period to another. In addition, MODFLOW-96 supports an option that makes it possible to determine the rate of induced streambed infiltration for each streambed reach.

MODFLOW grid design should take into consideration streambed reach widths and lengths. Streambed grid cells should be interconnected regardless of the configuration of streambed cells. Keep in mind that streambed-induced infiltration varies with the temperature of the surface water. A decline in the temperature of the surface water of 1° F will decrease the streambed hydraulic conductivity about 1.5% (Rorabaugh, 1956, pp. 101–169) and vice versa. Thus, the calculated streambed leakance must be assigned to a particular surface water temperature. During periods of high streamflow, the streambed is scoured and the streambed vertical hydraulic conductivity increases (see Nortz et al., 1994). During periods of low streamflow, fine materials are deposited in the streambed and the streambed leakance decreases (see Norris, 1983a, 1983b).

Streambed partial penetration can be simulated by dividing the aquifer thickness into several layers and assigning streambed data only to the upper layers. Partially penetrating wells can be simulated by assigning pumping rates to appropriate layers. Be aware that the cone of depression may expand only partway across the streambed if its streambed vertical hydraulic conductivity is high or the streambed is wide. If the streambed vertical hydraulic conductivity is low or the streambed is narrow, the cone of depression may expand across and beyond the streambed.

When heads in the aquifer are lowered below the streambed and underlying materials become unsaturated, aquifer heads calculated by MODFLOW are lower than actual and streambed leakance calculated with MODFLOW is less than actual because negative pressure heads in unsaturated materials are neglected (see Peterson, 1989, pp. 899–927). During the short period when unsaturated conditions prevail, aquifer head changes continue to affect the induced infiltration rate even though heads are below the streambed. The induced streambed infiltration rate continues to increase until further reduction in the moisture content in the unsaturated zone beneath the streambed creates a condition in which unsaturated materials are incapable of conveying water at the same rate as streamflow losses.

Osman and Bruen (2002) and Fox (2003) present methods for incorporating unsaturated flow into the MODFLOW River package.

DIMENSIONLESS TIME DRAWDOWN

Time-drawdown values calculated with MODFLOW can be converted to corresponding dimensionless time-drawdown values for selected type curve trace argument values so that the single plot interactive type curve matching technique can be applied to aquifer test data. Type curve trace values for uniform aquifer parameters and uncomplicated aquifer conditions can be calculated with MODFLOW equally as well as with analytical mathematical models. In addition, type curve trace values for nonuniform aquifer parameters or complicated aquifer conditions can be calculated with MODFLOW as illustrated in Figure 5.1 to Figure 5.3.

Dimensionless time drawdown is defined as follows:

$$\text{Dimensionless Drawdown} = 4\pi T_{av}s/Q \tag{5.29}$$

$$\text{Dimensionless Time} = 4T_{av}t/(r^2S_{av}) \tag{5.30}$$

where T_{av} is the average aquifer transmissivity within the pumped well cone of depression, s is the drawdown, Q is the pump discharge rate, t is the elapsed time, r is the distance from the pumped well, and S_{av} is the average aquifer storativity within the pumped well cone of depression.

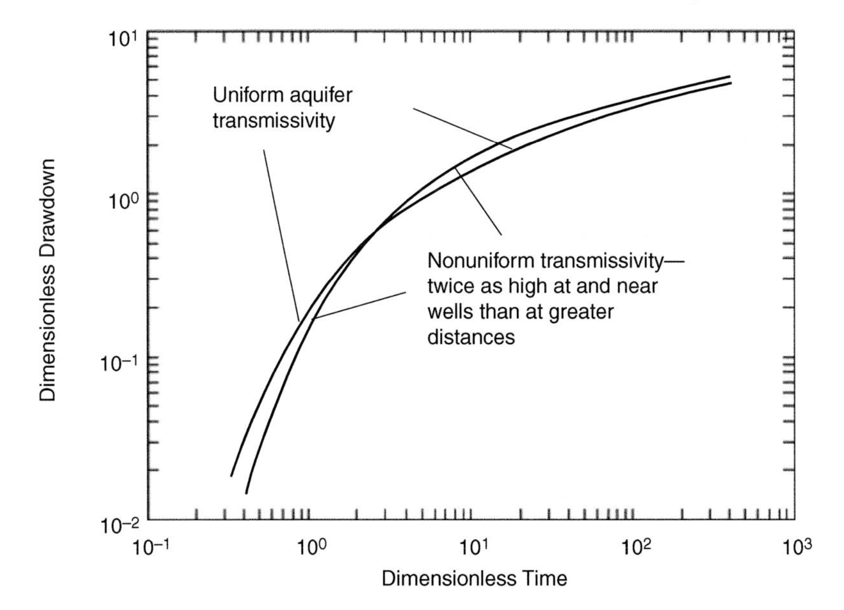

FIGURE 5.1 Graph showing nonuniform transmissivity effects on pumping test type curve values.

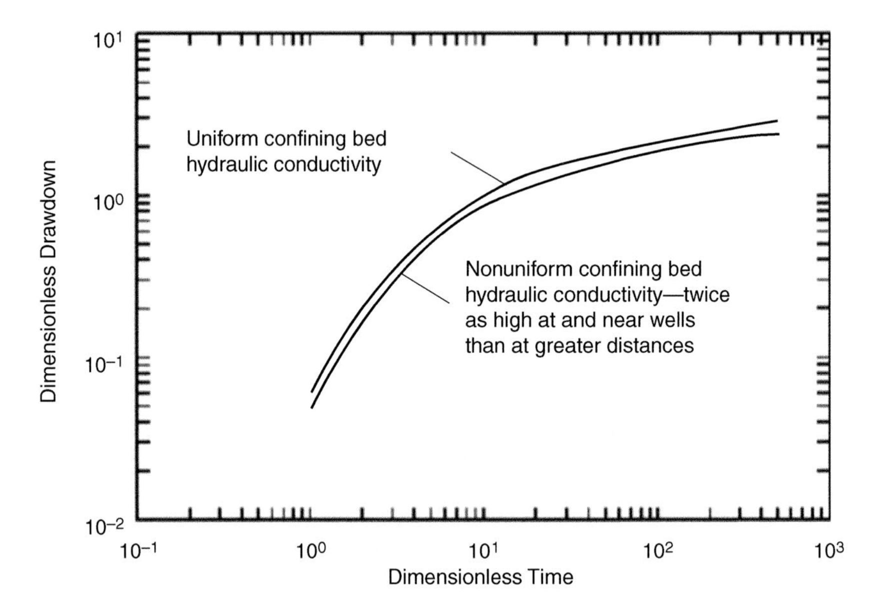

FIGURE 5.2 Graph showing nonuniform confining unit vertical hydraulic conductivity effects on pumping test type curve values.

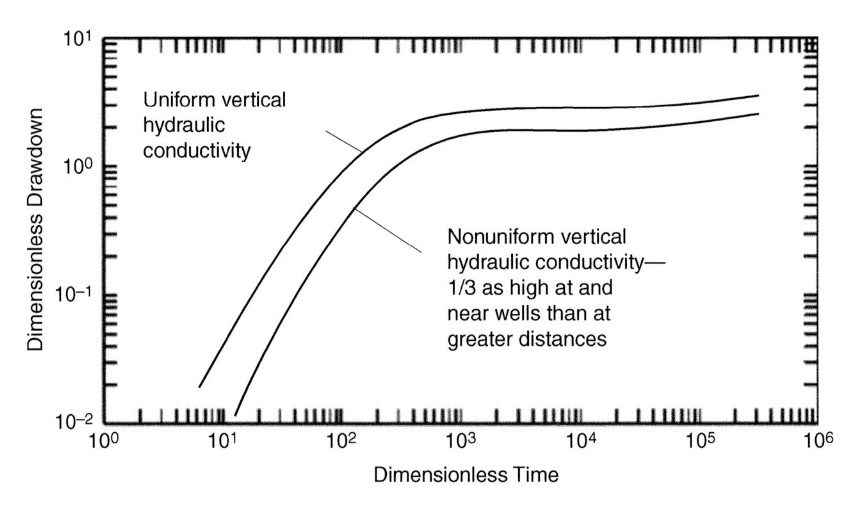

FIGURE 5.3 Graph showing nonuniform aquifer vertical hydraulic conductivity effects on pumping test type curve values.

Dimensionless time-drawdown values are associated with one of the following type curve trace arguments depending upon aquifer conditions:

Confined nonleaky aquifer:

$$\alpha = r_w^2\, S_{av}/(r_c^2 - r_p^2) \tag{5.31}$$

Confined leaky aquifer:

$$\tau = r/4[K'S'/(b'TS_{av})]^{0.5} \quad (5.32)$$

Confined fissure and block aquifer:

$$r_D = [r/(b'_b/z)](K'_b/K_f)^{0.5} \quad (5.33)$$

Unconfined aquifer:

$$\beta = (r^2K_v)/(b^2K_h) \quad (5.34)$$

where r_w is the pumped well effective radius, S_{av} is average aquifer storativity within the pumped well cone of depression, r_c is the pumped well casing radius, r_p is the pump pipe radius, K' is the confining unit or block vertical hydraulic conductivity, S' is the confining unit or block storativity, b' is the confining unit or block thickness, K_v is the aquifer vertical hydraulic conductivity, b is the aquifer thickness, K_h is the aquifer horizontal hydraulic conductivity, b'_b is the average block thickness between fissure zones, K'_b is the block vertical hydraulic conductivity, and K_f is the fissure horizontal hydraulic conductivity.

Dimensionless time-drawdown values calculated with MODFLOW can be used with external paper graphs or graphics software to analyze pumping test data.

6 Evaluation

An evaluation of the reliability and precision of calculated aquifer characteristic values is an important part of aquifer test modeling. The reliability and precision of pumping test results depend partly on the number, location, and penetration of observation wells. Reliability and precision generally increase with the number of observation wells up to about a dozen observation wells. Reliability and precision are diminished if the observation wells are not located appropriately with respect to aquifer hydrogeological characteristics such as trends in aquifer parameter values, heterogeneities, discontinuities, and boundaries. Reliability and precision are diminished if the observation well penetrations are not in tune with aquifer stratification conditions.

Reliability and precision are diminished if early time drawdowns are not measured or if the pumping test duration is not long enough to enable the complete analysis of the effects of any boundary, confining unit leakage, or delayed drainage at the water table.

Reliability and precision generally are increased by interactive calibration or automatic parameter estimation. However, it must be realized that interactive calibration programs and automatic parameter estimation programs, such as PEST, can only optimize aquifer parameter values based on aquifer test modeler-specified conceptual and mathematical models and ranges of adjustable parameter values. Optimizing programs cannot make judgments concerning the validity of the conceptual and mathematical models or ranges of adjustable parameter values. These judgments are reserved for the aquifer test modeler.

Aquifer test analysis results tend to be approximate and nonunique because test facilities are usually limited, test conditions are usually not ideal, field measurements are usually limited in accuracy and quantity (Stallman, 1971, p. 11), aquifer test conceptual models and equations seldom completely simulate reality, and observed time-drawdown or head values can be duplicated with more than one combination of aquifer parameter values and boundary conditions.

Pumping test analysis result precision depends partly on the degree of aquifer complexity. Calculated hydraulic conductivity values are often within 15% of actual hydraulic conductivity values and calculated storativity and specific yield values are often within 30% of actual storativity and specific yield values (see Walton, 1996, p. 91).

Calculated aquifer horizontal hydraulic conductivity and confined storativities values are mean (integrated) values for the aquifer test domain. The pumping test domain height is the aquifer thickness and the pumping test domain radius can be 1000 ft or more under confined nonleaky, leaky, or fissure and block aquifer conditions. The pumping test domain height is the aquifer thickness and the pumping test domain radius can be 500 ft or more under unconfined (water table) aquifer conditions. The slug test domain height is the aquifer thickness and the slug test domain radius can be 20 ft or more.

Under unconfined conditions, calculated specific yield values are mean values for the portion of the aquifer test domain between the initial water table and the upper surface of the cone of depression where the drainage of pores occurs, not the portion of the aquifer test domain below the upper surface of the cone of depression where confined conditions exist. The drained aquifer test domain height is the drawdown at any particular site that can be thin in large portions of the cone of depression and consist largely of fine-grained materials such as silt and clay. A drained aquifer test domain height cross section can be created with an appropriate analytical mathematical model. Be aware that the validity of calculated specific yield is controversial (Van der Kamp, 1985; Nwankwor et al., 1984, 1992; Akindunni and Gillham, 1992; Narasimhan and Zhu, 1993; and Moench et al., 2001).

Calculated aquifer vertical hydraulic conductivity and confined storativity values with partially penetrating wells are mean (integrated) values for only the portion of the aquifer test domain within which vertical components of flow occur. That portion of the aquifer test domain can have a height equal to the aquifer thickness and a radius of 100 ft or more.

Calculated aquifer parameter values for pumping and slug tests may differ by an order of magnitude because of differences between pumping and slug test domains. For this reason, if heterogeneity issues are of importance, it may be wise to conduct both pumping and slug tests.

Under confined leaky conditions, calculated confining unit vertical hydraulic conductivity and storativity values pertain to the portion of the confining unit within the vertical extent of the cone of depression, not necessarily the entire confining unit thickness. Usually, the cone of depression reaches the confining unit top within a very short time. However, the cone of depression may not reach the confining unit top during a pumping test if the confining unit is very thick and the confining unit vertical hydraulic conductivity is very low. The vertical extent of the cone of depression can be estimated with two-aquifer system with drawdown in the unpumped aquifer equations derived by Neuman and Witherspoon (1972) and presented by Moench and Ogata (1984).

Pumping test analysis results obtained with numerical mathematical modeling equations are not exact because of the approximations made in the numerical model (MODFLOW). These include use of a discrete rather than continuous spatial domain, use of a discrete rather than continuous time domain, and use of an iterative solution with convergence tolerance.

Slug test analysis result precision is affected by the following factors: well development, test design, and analysis procedures (Butler, 1998b). With proper development, the simulated aquifer hydraulic conductivity value should be within a factor of 2 of the actual aquifer hydraulic conductivity value. With improper development, the calculated aquifer hydraulic conductivity value can be in error by an order of magnitude or more. An inappropriate effective casing radius can lead to an aquifer hydraulic conductivity value that is in error by a factor of three to four. The Hvorslev and Bouwer and Rice method of analysis can lead to errors an order of magnitude or more. Three or more slug tests should be performed at each well to enable the effectiveness of well development and the viability of slug test theory to be evaluated at each well (Butler, 1998b, p. 27). The initial displacement and the direction of flow should be varied in repeat tests.

Conventional pumping test results yield little information concerning local changes in hydraulic conductivity (aquifer heterogeneity) because these changes have a disproportionately small impact on time-drawdown data. This is acceptable for water supply modeling, but not for mass transport modeling. In this case, some insight about aquifer heterogeneity can be obtained from data for several partially penetrating observation wells open at different depths and located close to the pumped well. Slug test results for scattered partially penetrating slug wells can provide additional heterogeneity information.

7 Sample Data File Sets

Five sample data file sets are provided herein to assist the reader in becoming familiar with WTAQ and MODFLOW-96 aquifer test modeling input and output data file contents with confined nonleaky and unconfined aquifer conditions. Conceptual models consisting of abbreviated descriptions of aquifer test facilities, aquifer test data, and aquifer parameter values are provided together with selected sample data file sets.

SAMPLE DATA FILE SET 1

Sample Data File Set 1 illustrates WTAQ data file sets generated during the calculation of pumping test type curve and time-drawdown values for a nonleaky aquifer. The WTAQ input data file and results and plot file contents are displayed after the conceptual model is briefly described.

The conceptual model pumping test facilities consist of a pumped well and one observation well. The pumping well casing and screen are both 10 in. in diameter (0.417 ft radius) and the observation well casing and screen are 4 in. in diameter (0.167 ft radius). The pumped well effective radius is 0.417 ft. The pump pipe diameter is 4 in. in diameter (0.167 ft radius). The pumped wellbore skin is assumed to be negligble. The distance between the pumped and observation well is 810 ft.

The pumped and observation wells both fully penetrate a sand and gravel aquifer that is 18 ft in thickness and occurs between the depths of 282 and 300 ft. The pumped well has wellbore storage. Shale underlies the aquifer and 282 ft of clay with a very low hydraulic conductivity (can be considered an impermeable unit for short periods of time) overlies the aquifer. The pumping rate is 220 gal/min (42,353 ft^3/day) and the pumping period duration is 8 h (0.347 day). Aquifer parameter values are transmissivity = 1318 ft^2/day, horizontal hydraulic conductivity = 73.22 ft/day, vertical hydraulic conductivity = 7.322 ft/day, and storativity = 0.00002.

TYPE CURVE FORMAT DATA-INPUT FILE

```
Sample Data File Set 1
TYPE CURVE
CONFINED
1.800E+01 1.000E-01 0.0
0 0
1.0E09
1.000E+06 7 8
0 1.000E-08 200 0 8
```

```
1 1
4.170E-01 0.000E+00 1.800E+01 2.098E+04 0.000E+00
1
1 0
8.100E+02 0.000E+00 1.800E+01 0.0 0.0
```

TYPE CURVE FORMAT RESULTS OUTPUT FILE

```
*****************************************************************
*                                                               *
*              **** U.S. GEOLOGICAL SURVEY ****                 *
*                                                               *
*               **** WTAQ: PROGRAM OUTPUT ****                  *
*                                                               *
*          COMPUTER PROGRAM FOR CALCULATING DRAWDOWN            *
*                                                               *
*          IN A CONFINED OR WATER-TABLE AQUIFER WITH            *
*                                                               *
*             AXIAL-SYMMETRIC FLOW TO A FINITE- OR              *
*                                                               *
*              INFINITESIMAL-DIAMETER PUMPED WELL               *
*                                                               *
*                   VERSION 1.0, 10/01/99                       *
*                                                               *
*****************************************************************
Sample Data File Set 1

TYPE-CURVE FORMAT                CONFINED AQUIFER

        *** AQUIFER HYDRAULIC PROPERTIES ***
SATURATED THICKNESS (BB):        0.180D+02 (units of length)
RATIO OF VERTICAL TO HORIZONTAL
 HYDRAULIC CONDUCTIVITY (XKD):   0.100D+00 (dimensionless)

         *** PROGRAM SOLUTION VARIABLES  ***
 LARGEST VALUE        NUMBER OF      DRAWDOWN CALCULATIONS
OF TIME (TDLAST)   LOG CYCLES (NLC)   PER LOG CYCLE (NOX)
----------------- ---------------- ---------------------
    0.100D+07            7                    8

    RERRNR       RERRSUM     NMAX   NTMS   NS
  ---------    ---------     ----   ----   --
  0.000D+00    0.100D-07     200      0    8
```

```
 *** PUMPED-WELL CHARACTERISTICS AND CALCULATED DRAWDOWN ***

  WELL-DIAMETER TYPE (IPWD):  1 (finite diameter)
  SCREENED INTERVAL (IPWS):   1 (fully penetrating)

                SCREENED   INTERVAL   WELL BORE   WELL BORE
  WELL RADIUS     ZPD         ZPL      STORAGE      SKIN
  -----------   --------   --------   ---------   ---------
   0.42D+00     0.00D+00   0.18D+02    0.21D+05    0.00D+00

  BETAW = 0.537D-04

  DIMENSIONLESS TIME    DIMENSIONLESS DRAWDOWN
       (TDRDSQ)                  (HD)
  -------------------   ----------------------
      0.1000D+00             0.0000D+00
      0.1334D+00             0.0000D+00
                      .
                      .
      0.7499D+06             0.1374D+02
      0.1000D+07             0.1423D+02

** OBSERVATION-WELL CHARACTERISTICS AND CALCULATED DRAWDOWN **

          **** OBSERVATION WELL OR PIEZOMETER 1 ****

  FULLY PENETRATING OBSERVATION WELL

  DISTANCE FROM
    CENTER OF                                  DELAYED RESPONSE
   PUMPED WELL        Z1          Z2                FACTOR
  -------------   ---------   ---------    ----------------
    0.810D+03     0.000D+00   0.180D+02        0.000D+00

  DRAWDOWN CALCULATED FOR BETA = 0.202D+03

   DIMENSIONLESS TIME   DIMENSIONLESS DRAWDOWN
        (TDRDSQ)                 (HD)
   ------------------   ----------------------
       0.1000D+00            0.0000D+00
       0.1334D+00            0.3087D-01
                      .
                      .
       0.7499D+06            0.1434D+02
       0.1000D+07            0.1462D+02
```

TYPE CURVE FORMAT PLOT OUTPUT FILE

```
  TDRDSQ      HDT      TDYRDSQ      HDTY      HDPW        HDOB1
0.100E+00 0.248E-01 0.100E+00 0.248E-01 0.000E+00 0.000E+00
0.133E+00 0.587E-01 0.133E+00 0.587E-01 0.000E+00 0.309E-01
                                    .
                                    .
0.750E+06 0.143E+02 0.750E+06 0.143E+02 0.137E+02 0.143E+02
0.100E+07 0.146E+02 0.100E+07 0.146E+02 0.142E+02 0.146E+02
```

DIMENSIONAL FORMAT DATA-INPUT FILE

```
Sample Data File Set 1
DIMENSIONAL
CONFINED
1.800E+01 7.322E+01 7.322E+00 1.111E-06 0.0
0 0
1.0E09
1 1
0.0 0 0
0.0 1.000E-08 200 0 8
1 1
4.235E+04 4.170E-01 3.820E-01 0.000E+00 1.800E+01 0.000E+00
0 0
1
1 0
8.100E+02 0.000E+00 1.800E+01 0.0 0.0 0.0
22 1
2.100E-03 3.000E-01
3.500E-03 7.000E-01
       .
       .
2.640E-01 1.020E+01
3.473E-01 1.090E+01
```

DIMENSIONAL FORMAT RESULTS FILE

```
***************************************************************
*                                                             *
*              **** U.S. GEOLOGICAL SURVEY ****               *
*                                                             *
*               **** WTAQ: PROGRAM OUTPUT ****                *
*                                                             *
*          COMPUTER PROGRAM FOR CALCULATING DRAWDOWN          *
*                                                             *
*          IN A CONFINED OR WATER-TABLE AQUIFER WITH          *
```

```
*                                                              *
*            AXIAL-SYMMETRIC FLOW TO A FINITE- OR              *
*                                                              *
*             INFINITESIMAL-DIAMETER PUMPED WELL               *
*                                                              *
*                   VERSION 1.0, 10/01/99                      *
*                                                              *
****************************************************************
Sample Data File Set 1

DIMENSIONAL FORMAT       CONFINED AQUIFER

             *** AQUIFER HYDRAULIC PROPERTIES ***

SATURATED THICKNESS (BB): 0.180D+02 (units of length)
HORIZONTAL HYDRAULIC
 CONDUCTIVITY (HKR):      0.732D+02 (units of length per time)
VERTICAL HYDRAULIC
 CONDUCTIVITY (HKZ):      0.732D+01 (units of length per time)
RATIO OF VERTICAL TO HORIZONTAL
 HYDRAULIC CONDUCTIVITY (XKD): 0.100D+00 (dimensionless)
CALCULATED TRANSMISSIVITY:     0.132D+04 (units of length
                                          squared per time)
SPECIFIC STORAGE (SS):    0.111D-05 (units of inverse length)
CALCULATED STORATIVITY:   0.200D-04 (dimensionless)

              *** PROGRAM SOLUTION VARIABLES  ***

   USER-SPECIFIED TIMES; MEASURED DRAWDOWN DATA SPECIFIED

    RERRNR       RERRSUM     NMAX   NTMS   NS
  ---------    ---------    ----   ----   --
  0.000D+00    0.100D-07     200      0    8

 *** PUMPED-WELL CHARACTERISTICS AND CALCULATED DRAWDOWN ***

  WELL-DIAMETER TYPE (IPWD): 1 (finite diameter)
  SCREENED INTERVAL (IPWS):  1 (fully penetrating)
  PUMPING RATE OF WELL (QQ): 0.424D+05 (cubic length per time)

                SCREENED   INTERVAL   WELL BORE   WELL BORE
  WELL RADIUS     ZPD        ZPL       STORAGE       SKIN
  -----------   --------   --------   ---------   ---------
   0.42D+00     0.00D+00   0.18D+02   0.21D+05    0.00D+00

  BETAW =    0.537D-04
```

```
** OBSERVATION-WELL CHARACTERISTICS AND CALCULATED DRAWDOWN **

        ****   OBSERVATION WELL OR PIEZOMETER  1  ****

   FULLY PENETRATING OBSERVATION WELL

  DISTANCE FROM
    CENTER OF                                 DELAYED RESPONSE
   PUMPED WELL        Z1           Z2              FACTOR
  -------------    ---------    ---------    ----------------
    0.810D+03      0.000D+00    0.180D+02         0.000D+00

  DRAWDOWN CALCULATED FOR BETA = 0.202D+03

                                                 RELATIVE
                  MEASURED      CALCULATED         ERROR
   TIME           DRAWDOWN      DRAWDOWN         (PERCENT)
  ----------     ----------    ----------       ----------
  0.2100D-02     0.3000D+00    0.2964D+00       0.1194D+01
  0.3500D-02     0.7000D+00    0.7947D+00      -0.1353D+02
                           .
                           .
  0.2640D+00     0.1020D+02    0.1047D+02      -0.2626D+01
  0.3473D+00     0.1090D+02    0.1116D+02      -0.2429D+01
```

DIMENSIONAL FORMAT PLOT OUTPUT FILE

```
      ****   OBSERVATION WELL OR PIEZOMETER  1  ****

    TIME          MEASDD        CALCDD          RELERR
  0.2100E-02     0.3000E+00    0.2964E+00       0.1194E+01
  0.3500E-02     0.7000E+00    0.7947E+00      -0.1353E+02
                           .
                           .
  0.1806E+00     0.9200E+01    0.9506E+01      -0.3322E+01
  0.2223E+00     0.9700E+01    0.1003E+02      -0.3420E+01
  0.2640E+00     0.1020E+02    0.1047E+02      -0.2626E+01
  0.3473E+00     0.1090E+02    0.1116E+02      -0.2429E+01
```

SAMPLE DATA FILE SET 2

Sample Data File Set 2 illustrates WTAQ and PEST data sets generated during automatic calibration of pumping test time-drawdown values for an unconfined aquifer. The WTAQ input data file, results file, and plot file contents and PEST input files and output data files are displayed after the conceptual model is briefly described.

Conceptual modeling pumping test facilities consist of a partially penetrating pumped well and four partial penetrating observation piezometers in an unconfined sand and gravel aquifer. The effective radius of the pumped well is 0.1 m. The effective casing radius of the pumped well is 0.1 m. The pumped well screen length is 5.0 m. The depth between the initial water table and the pumped well screen base is 10.0 m. The depth between the initial water table and the pumped well screen top is 5.0 m. The pumped wellbore skin is assumed to be negligible. The pumped well has a finite diameter. The constant pumping rate is 2.0E-3 m^3/sec.

The radial distances from the pumped well to the observation piezometers 1, 2, 3, and 4 are 3.16 m, 31.6 m, 3.16 m, and 31.6 m, respectively. The depths below the initial water table to the centers of observation piezometers 1, 2, 3, and 4 are 7.5 m, 7.5 m, 1.0 m, and 1.0 m, respectively. The observation piezometer radii are each 0.05 m. It is assumed that the effects of delayed drawdown response at the observation piezometers have been effectively eliminated with hydraulic packers.

The aquifer is 10.0 m thick. The drainage at the water table is instantaneous and the single drainage constant is 1.0E9 1/sec. Data for both the Type Curve and the Dimensional formats is to be calculated. Optimized aquifer parameters are horizontal hydraulic conductivity, vertical hydraulic conductivity, storativity, and specific yield of the aquifer are 1.01E-4 m/sec, 5.14E-5 m/sec, 3.99E-4, and 0.2, respectively.

WTAQ INPUT DATA FILE

```
Sample Data File Set 2
DIMENSIONAL
WATER TABLE
1.000E+01   1.0079829E-04   5.2020815E-05   3.2263170E-05
  2.0028355E-01
0 0
1.0E09
1 1
0.0 0 0
1.000E-10 0.0 0 30 8
0 1
2.000E-03 1.000E-01 1.000E-01 5.000E+00 1.000E+01 0.000E+00
14 1
9.300E+00 5.100E-01
2.000E+01 9.700E-01
         .
         .
9.280E+04 2.980E+00
2.000E+05 3.090E+00
4
```

```
2 0
3.16 0.0 0.0 7.500E+00 0.0 0.0
14 1
9.300E+00 9.000E-02
2.000E+01 2.040E-01
        .
        .
9.280E+04 9.420E-01
2.000E+05 1.052E+00
2 0
31.6 0.0 0.0 7.500E+00 0.0 0.0
14 1
9.300E+00 0.000E+00
2.000E+01 5.000E-04
        .
        .
9.280E+04 9.830E-02
2.000E+05 1.722E-01
2 0
3.16 0.0 0.0 1.000E+00 0.0 0.0
14 1
9.300E+00 7.000E-03
2.000E+01 1.900E-02
        .
        .
9.280E+04 4.878E-01
2.000E+05 6.126E-01
2 0
31.6 0.0 0.0 1.000E+00 0.0 0.0
14 1
9.300E+00 0.000E+00
2.000E+01 1.000E-04
4.310E+01 5.000E-04
        .
        .
9.280E+04 8.400E-02
2.000E+05 1.620E-01
```

WTAQ REDIRECT FILE

```
c:\sample2\wtaqred.txt
c:\sample2\wtaqres.txt
c:\sample2\wtaqplot.txt
```

WTAQ RESULTS OUTPUT FILE

```
************************************************************
*              **** U.S. GEOLOGICAL SURVEY ****            *
*                                                          *
*               **** WTAQ: PROGRAM OUTPUT ****             *
*                                                          *
*         COMPUTER PROGRAM FOR CALCULATING DRAWDOWN        *
*                                                          *
*         IN A CONFINED OR WATER-TABLE AQUIFER WITH        *
*                                                          *
*            AXIAL-SYMMETRIC FLOW TO A FINITE- OR          *
*                                                          *
*             INFINITESIMAL-DIAMETER PUMPED WELL           *
*                                                          *
*                   VERSION 1.0, 10/01/99                  *
************************************************************

Sample Data File Set 2

DIMENSIONAL FORMAT        WATER-TABLE AQUIFER

              *** AQUIFER HYDRAULIC PROPERTIES ***

SATURATED THICKNESS (BB): 0.100D+02 (units of length)
HORIZONTAL HYDRAULIC
 CONDUCTIVITY (HKR):      0.101D-03 (units of length per time)
VERTICAL HYDRAULIC
 CONDUCTIVITY (HKZ):      0.520D-04 (units of length per time)
RATIO OF VERTICAL TO HORIZONTAL
 HYDRAULIC CONDUCTIVITY (XKD): 0.516D+00 (dimensionless)
CALCULATED TRANSMISSIVITY:     0.101D-02 (units of length
                                          squared per time)
SPECIFIC STORAGE (SS):   0.323D-04 (units of inverse length)
SPECIFIC YIELD (SY):     0.200D+00 (dimensionless)
CALCULATED STORATIVITY:  0.323D-03 (dimensionless)
RATIO OF STORATIVITY TO
 SPECIFIC YIELD (SIGMA): 0.161D-02 (dimensionless)
DRAINAGE AT WATER TABLE (IDRA): 0 (instantaneous)

           *** PROGRAM SOLUTION VARIABLES  ***

USER-SPECIFIED TIMES; MEASURED DRAWDOWN DATA SPECIFIED

    RERRNR       RERRSUM     NMAX   NTMS   NS
  ---------    ---------     ----   ----   --
  0.100D-09    0.000D+00       0     30    8
```

```
  *** PUMPED-WELL CHARACTERISTICS AND CALCULATED DRAWDOWN ***

WELL-DIAMETER TYPE (IPWD):  1 (finite diameter)
SCREENED INTERVAL (IPWS):   0 (partially penetrating)
PUMPING RATE OF WELL (QQ):  0.200D-02 (cubic length per time)

                SCREENED   INTERVAL   WELL BORE   WELL BORE
  WELL RADIUS      ZPD        ZPL      STORAGE       SKIN
  -----------   --------   --------   ---------   ---------
    0.10D+00    0.50D+01   0.10D+02    0.31D+04    0.00D+00

BETAW =    0.516D-04
                                                     RELATIVE
                    MEASURED      CALCULATED          ERROR
   TIME             DRAWDOWN       DRAWDOWN         (PERCENT)
  ----------       ----------     ----------       ----------
  0.9300D+01    0.5100D+00     0.5188D+00       -0.1727D+01
  0.2000D+02    0.9700D+00     0.9857D+00       -0.1614D+01
                                      .
                                      .
  0.9280D+05     0.2980D+01     0.2979D+01       0.2535D-01
  0.2000D+06     0.3090D+01     0.3090D+01      -0.1557D-01

** OBSERVATION-WELL CHARACTERISTICS AND CALCULATED DRAWDOWN **

         ****   OBSERVATION WELL OR PIEZOMETER  1   ****

  OBSERVATION PIEZOMETER

   DISTANCE FROM
     CENTER OF                 DELAYED RESPONSE
    PUMPED WELL       ZP            FACTOR
   -------------   ---------   ----------------
     0.316D+01     0.750D+01       0.000D+00

  DRAWDOWN CALCULATED FOR BETA = 0.515D-01

                                                     RELATIVE
                    MEASURED      CALCULATED          ERROR
   TIME             DRAWDOWN       DRAWDOWN         (PERCENT)
  ----------       ----------     ----------       ----------
  0.9300D+01     0.9000D-01     0.7653D-01       0.1497D+02
  0.2000D+02     0.2040D+00     0.1884D+00       0.7649D+01
                                       .
                                       .
```

```
 0.9280D+05       0.9420D+00      0.9462D+00     -0.4448D+00
 0.2000D+06       0.1052D+01      0.1057D+01     -0.4368D+00

     ****   OBSERVATION WELL OR PIEZOMETER  2  ****

  OBSERVATION PIEZOMETER

  DISTANCE FROM
    CENTER OF                   DELAYED RESPONSE
   PUMPED WELL         ZP            FACTOR
  -------------    ---------   ----------------
    0.316D+02      0.750D+01       0.000D+00

DRAWDOWN CALCULATED FOR BETA = 0.515D+01

                                                   RELATIVE
                   MEASURED      CALCULATED         ERROR
  TIME             DRAWDOWN       DRAWDOWN        (PERCENT)
 ----------       ----------     ----------      ----------
 0.9300D+01       0.0000D+00     0.0000D+00         ****
 0.2000D+02       0.5000D-03     0.0000D+00      0.1000D+03
                             .
                             .
 0.9280D+05       0.9830D-01     0.9849D-01     -0.1934D+00
 0.2000D+06       0.1722D+00     0.1740D+00     -0.1060D+01

     ****   OBSERVATION WELL OR PIEZOMETER  3  ****

  OBSERVATION PIEZOMETER

  DISTANCE FROM
    CENTER OF                   DELAYED RESPONSE
   PUMPED WELL         ZP            FACTOR
  -------------    ---------   ----------------
    0.316D+01      0.100D+01       0.000D+00

  DRAWDOWN CALCULATED FOR BETA = 0.515D-01

                                                   RELATIVE
                   MEASURED      CALCULATED         ERROR
  TIME             DRAWDOWN       DRAWDOWN        (PERCENT)
 ----------       ----------     ----------      ----------
 0.9300D+01       0.7000D-02     0.4515D-02      0.3550D+02
 0.2000D+02       0.1900D-01     0.1503D-01      0.2089D+02
                             .
                             .
```

```
0.9280D+05     0.4878D+00     0.4892D+00    -0.2799D+00
0.2000D+06     0.6126D+00     0.6129D+00    -0.4483D-01

      ****   OBSERVATION WELL OR PIEZOMETER  4  ****

 OBSERVATION PIEZOMETER

 DISTANCE FROM
   CENTER OF                    DELAYED RESPONSE
  PUMPED WELL        ZP              FACTOR
 -------------   ---------   ----------------
   0.316D+02     0.100D+01        0.000D+00

DRAWDOWN CALCULATED FOR BETA = 0.515D+01

                                               RELATIVE
                  MEASURED       CALCULATED      ERROR
 TIME             DRAWDOWN        DRAWDOWN     (PERCENT)
----------       ----------      ----------    ----------
0.9300D+01     0.0000D+00     0.0000D+00        ****
0.2000D+02     0.1000D-03     0.0000D+00     0.1000D+03
                                  .
                                  .
0.9280D+05     0.8400D-01     0.8517D-01    -0.1394D+01
0.2000D+06     0.1620D+00     0.1650D+00    -0.1876D+01
```

WTAQ PLOT OUTPUT FILE

```
                ****   PUMPED WELL   ****

  TIME           MEASDD          CALCDD          RELERR
0.9300E+01     0.5100E+00     0.5188E+00    -0.1727E+01
0.2000E+02     0.9700E+00     0.9857E+00    -0.1614E+01
                                  .
                                  .
0.9280E+05     0.2980E+01     0.2979E+01     0.2535E-01
0.2000E+06     0.3090E+01     0.3090E+01    -0.1557E-01

     ****   OBSERVATION WELL OR PIEZOMETER  1  ****

  TIME           MEASDD          CALCDD          RELERR
0.9300E+01     0.9000E-01     0.7653E-01     0.1497E+02
0.2000E+02     0.2040E+00     0.1884E+00     0.7649E+01
                                  .
                                  .
0.9280E+05     0.9420E+00     0.9462E+00    -0.4448E+00
0.2000E+06     0.1052E+01     0.1057E+01    -0.4368E+00
```

```
       ****   OBSERVATION WELL OR PIEZOMETER  2  ****

     TIME          MEASDD          CALCDD          RELERR
  0.9300E+01     0.0000E+00     0.0000E+00         ****
  0.2000E+02     0.5000E-03     0.0000E+00     0.1000E+03
                              .
                              .
  0.9280E+05     0.9830E-01     0.9849E-01    -0.1934E+00
  0.2000E+06     0.1722E+00     0.1740E+00    -0.1060E+01

       ****   OBSERVATION WELL OR PIEZOMETER  3  ****

     TIME          MEASDD          CALCDD          RELERR
  0.9300E+01     0.7000E-02     0.4515E-02     0.3550E+02
  0.2000E+02     0.1900E-01     0.1503E-01     0.2089E+02
                              .
                              .
  0.9280E+05     0.4878E+00     0.4892E+00    -0.2799E+00
  0.2000E+06     0.6126E+00     0.6129E+00    -0.4483E-01

       ****   OBSERVATION WELL OR PIEZOMETER  4  ****

     TIME          MEASDD          CALCDD          RELERR
  0.9300E+01     0.0000E+00     0.0000E+00         ****
  0.2000E+02     0.1000E-03     0.0000E+00     0.1000E+03
                              .
                              .
  0.9280E+05     0.8400E-01     0.8517E-01    -0.1394E+01
  0.2000E+06     0.1620E+00     0.1650E+00    -0.1876E+01
```

PEST TEMPLATE FILE

```
ptf #
Sample Data File 2
DIMENSIONAL
WATER TABLE
1.000E+01 #hhc1     # #vhc1     # #sto1     # #Spy1     #
0 0
1.0E09
1 1
0.0 0 0
1.000E-10 0.0 0 30 8
0 1
2.000E-03 1.000E-01 1.000E-01 5.000E+00 1.000E+01 0.000E+00
14 1
```

```
9.300E+00 5.100E-01
2.000E+01 9.700E-01
          .
          .
9.280E+04 2.980E+00
2.000E+05 3.090E+00
4
2 0
3.16 0.0 0.0 7.500E+00 0.0 0.0
14 1
9.300E+00 9.000E-02
2.000E+01 2.040E-01
          .
          .
9.280E+04 9.420E-01
2.000E+05 1.052E+00
2 0
31.6 0.0 0.0 7.500E+00 0.0 0.0
14 1
9.300E+00 0.000E+00
2.000E+01 5.000E-04
          .
          .
9.280E+04 9.830E-02
2.000E+05 1.722E-01
2 0
3.16 0.0 0.0 1.000E+00 0.0 0.0
14 1
9.300E+00 7.000E-03
2.000E+01 1.900E-02
          .
          .
9.280E+04 4.878E-01
2.000E+05 6.126E-01
2 0
31.6 0.0 0.0 1.000E+00 0.0 0.0
14 1
9.300E+00 0.000E+00
2.000E+01 1.000E-04
          .
          .
9.280E+04 8.400E-02
2.000E+05 1.620E-01
```

PEST INSTRUCTION FILE

```
pif @
l6 w w w !pwc1!
l1 w w w !pwc2!
l1 w w w !pwc3!
       .
       .
l1 w w w !pwc13!
l1 w w w !pwc14!
l6 w w w !owc11!
l1 w w w !owc12!
       .
       .
l1 w w w !owc113!
l1 w w w !owc114!
l6 w w w !owc21!
l1 w w w !owc22!
       .
       .
l1 w w w !owc213!
l1 w w w !owc214!
l6 w w w !owc31!
l1 w w w !owc32!
       .
       .
l1 w w w !owc313!
l1 w w w !owc314!
l6 w w w !owc41!
l1 w w w !owc42!
       .
       .
l1 w w w !owc413!
l1 w w w !owc414!
```

PEST CONTROL FILE

```
pcf
* control data
restart estimation
4 70 3 0 5
1 1 single point 1 0 0
5.0 2.0 0.3 0.01 10
5.0 5.0 0.001
0.1
```

```
30 0.01 4 4 0.01 3
1 1 1
* parameter groups
pgn1 relative 0.01 0.0 switch 1.0 parabolic
pgn2 relative 0.01 0.0 switch 1.0 parabolic
pgn3 relative 0.01 0.0 switch 1.0 parabolic
* parameter data
hhc1 log factor 1.0E-4 5.0E-5 1.5E-4 pgn1 1 0 1
vhc1 log factor 5.0E-5 1.0E-5 1.0E-4 pgn1 1 0 1
sto1 log factor 2.0E-5 1.0E-5 4.0E-5 pgn2 1 0 1
spy1 log factor 0.2 0.1 0.3 pgn3 1 0 1
* observation groups
ogn1
ogn2
ogn3
ogn4
ogn5
* observation data
pwc1 0.51 1.0 ogn1
pwc2 0.97 1.0 ogn1
         .
         .
pwc13 2.98 1.0 ogn1
pwc14 3.09 1.0 ogn1
owc11 0.09 1.0 ogn2
owc12 0.204 1.0 ogn2
         .
         .
owc113 0.942 1.0 ogn2
owc114 1.052 1.0 ogn2
owc21 0.0000 1.0 ogn3
owc22 0.0005 1.0 ogn3
         .
         .
owc213 0.0983 1.0 ogn3
owc214 0.172 1.0 ogn3
owc31 0.007 1.0 ogn4
owc32 0.019 1.0 ogn4
         .
         .
owc313 0.487 1.0 ogn4
owc314 0.612 1.0 ogn4
owc41 0.0000 1.0 ogn5
owc42 0.0001 1.0 ogn5
         .
         .
```

```
owc413 0.084 1.0 ogn5
owc414 0.162 1.0 ogn5
* model command line
c:\wtaq\wtaq.exe < c:\sample2\wtaqred.txt
* model input/output
c:\pest\temp.txt c:\wtaq\samp2inp.txt
c:\pest\ins.txt c:\wtaq\samp2plot.txt
```

PEST PARAMETER OUTPUT FILE

```
single point
  hhc1    1.0095641E-04       1.000000          0.000000
  vhc1    5.1409653E-05       1.000000          0.000000
  sto1    3.9813856E-05       1.000000          0.000000
  spy1    0.1982720           1.000000          0.000000
```

PEST RECORD OUTPUT FILE

```
          PEST RUN RECORD: CASE SAMPLE DATA FILE SET 2

PEST run mode:-

   Parameter estimation mode

Case dimensions:-

   Number of parameters                   :    4
   Number of adjustable parameters        :    4
   Number of parameter groups             :    3
   Number of observations                 :   70
   Number of prior estimates              :    0

Model command line(s):-

   c:\wtaq\wtaq.exe < c:\wtaq\sample2red.txt

Jacobian command line:-

   na

Model interface files:-

   Templates:
      c:\pest\samp2temp.txt
   for model input files:
      c:\wtaq\samp2inp.txt

   (Parameter values written using single precision protocol.)
   (Decimal point always included.)
```

```
   Instruction files:
      c:\pest\samp2ins.txt
   for reading model output files:
      c:\pest\samp2out.txt

PEST-to-model message file:-

   na

Derivatives calculation:-

Param  Increment  Increment   Increment  Forward    Multiplier  Method
group    type       low         bound    or central (central)   (central)
-------------------------------------------------------------------------
pgn1   relative   1.0000E-02   none       switch      1.000     parabolic
pgn2   relative   1.0000E-02   none       switch      1.000     parabolic
pgn3   relative   1.0000E-02   none       switch      1.000     parabolic

Parameter definitions:-

Name     Trans-      Change       Initial          Lower            Upper
         formation   limit        value            bound            bound
hhc1       log       factor   1.000000E-04    5.000000E-05    1.500000E-04
vhc1       log       factor   5.000000E-05    1.000000E-05    1.000000E-04
sto1       log       factor   2.000000E-05    1.000000E-05    4.000000E-05
spy1       log       factor   0.200000        0.100000        0.300000

Name        Group        Scale        Offset       Model command number
hhc1        pgn1        1.00000      0.00000              1
vhc1        pgn1        1.00000      0.00000              1
sto1        pgn2        1.00000      0.00000              1
spy1        pgn3        1.00000      0.00000              1

Prior information:-

   No prior information supplied

Observations:-

Observation name    Observation        Weight      Group
 pwc1               0.510000           1.000       ogn1
 pwc2               0.970000           1.000       ogn1
                                 .
                                 .
 pwc13               2.98000           1.000       ogn1
 pwc14               3.09000           1.000       ogn1
 owc11               9.000000E-02      1.000       ogn2
 owc12               0.204000          1.000       ogn2
                                 .
                                 .
 owc113              0.942000          1.000       ogn2
 owc114              1.05200           1.000       ogn2
```

```
 owc21               0.00000           1.000        ogn3
 owc22               5.000000E-04      1.000        ogn3
                          .
                          .
 owc213              9.830000E-02      1.000        ogn3
 owc214              0.172000          1.000        ogn3
 owc31               7.000000E-03      1.000        ogn4
 owc32               1.900000E-02      1.000        ogn4
                          .
                          .
 owc313              0.487000          1.000        ogn4
 owc314              0.612000          1.000        ogn4
 owc41               0.00000           1.000        ogn5
 owc42               1.000000E-04      1.000        ogn5
                          .
                          .
 owc413              8.400000E-02      1.000        ogn5
 owc414              0.162000          1.000        ogn5

Control settings:-

 Initial lambda                                        :  5.0000
 Lambda adjustment factor                              :  2.0000
 Sufficient new/old phi ratio per optimisation
   iteration                                           : 0.30000
 Limiting relative phi reduction between lambdas       : 1.00000E-02
 Maximum trial lambdas per iteration                   :  10

 Maximum  factor  parameter change (factor-
   limited changes)                                    :  5.0000
 Maximum relative parameter change (relative-
   limited changes)                                    :   na
 Fraction of initial parameter values used in computing
   change limit for near-zero parameters               : 1.00000E-03

 Relative phi reduction below which to begin use of
   central derivatives                                 : 0.10000

 Relative phi reduction indicating convergence         : 0.10000E-01
 Number of phi values required within this range       :   4
 Maximum number of consecutive failures
   to lower phi                                        :   4
 Minimal relative parameter change indicating
   convergence                                         : 0.10000E-01
```

```
 Number of consecutive iterations with minimal
   param change                                        :   3
 Maximum number of optimisation iterations             :  30

                          OPTIMISATION RECORD

INITIAL CONDITIONS:
Sum of squared weighted residuals (ie phi)              = 2.16294E-02
Contribution to phi from observation group "ogn1" = 1.86005E-02
Contribution to phi from observation group "ogn2" = 2.94933E-03
Contribution to phi from observation group "ogn3" = 1.06870E-05
Contribution to phi from observation group "ogn4" = 5.15230E-05
Contribution to phi from observation group "ogn5" = 1.73952E-05

      Current parameter values
      hhc1            1.000000E-04
      vhc1            5.000000E-05
      sto1            2.000000E-05
      spy1            0.200000

OPTIMISATION ITERATION NO.          :    1
   Model calls so far               :    1
Starting phi for this iteration                         : 2.16294E-02
Contribution to phi from observation group "ogn1" : 1.86005E-02
Contribution to phi from observation group "ogn2" : 2.94933E-03
Contribution to phi from observation group "ogn3" : 1.06870E-05
Contribution to phi from observation group "ogn4" : 5.15230E-05
Contribution to phi from observation group "ogn5" : 1.73952E-05

       Lambda =  5.0000      ----->
          Phi =  7.91036E-03  (  0.366 of starting phi)

       Lambda =  2.5000      ----->
          Phi =  7.67146E-03  (  0.355 of starting phi)

       Lambda =  1.2500      ----->
          Phi =  7.29862E-03  (  0.337 of starting phi)

       Lambda =  0.62500      ----->
          Phi =  6.83076E-03  (  0.316 of starting phi)

       Lambda =  0.31250      ----->
          Phi =  6.46900E-03  (  0.299 of starting phi)

   No more lambdas: phi is less than 0.3000 of starting phi
   Lowest phi this iteration:  6.46900E-03
```

```
  Current parameter values          Previous parameter values
     hhc1     1.011633E-04            hhc1     1.000000E-04
     vhc1     5.120363E-05            vhc1     5.000000E-05
     sto1     2.551447E-05            sto1     2.000000E-05
     spy1     0.195926                spy1     0.200000
  Maximum   factor change:  1.276     ["sto1"]
  Maximum relative change: 0.2757     ["sto1"]

OPTIMISATION ITERATION NO.        :    2
 Model calls so far               :   10
 Starting phi for this iteration:  6.46900E-03
 Contribution to phi from observation group "ogn1": 5.84529E-03
 Contribution to phi from observation group "ogn2": 4.83752E-04
 Contribution to phi from observation group "ogn3": 2.77649E-05
 Contribution to phi from observation group "ogn4": 6.93569E-05
 Contribution to phi from observation group "ogn5": 4.28386E-05

      Lambda =  0.15625     ----->
         Phi =  4.67748E-03  (  0.723 of starting phi)

      Lambda =  7.81250E-02 ----->
         Phi =  4.68367E-03  (  0.724 of starting phi)

      Lambda =  0.31250     ----->
         Phi =  4.71113E-03  (  0.728 of starting phi)

   No more lambdas: phi rising
   Lowest phi this iteration:  4.67748E-03

  Current parameter values          Previous parameter values
    hhc1   1.008327E-04               hhc1     1.011633E-04
    vhc1   5.165646E-05               vhc1     5.120363E-05
    sto1   3.359825E-05               sto1     2.551447E-05
    spy1   0.200313                   spy1     0.195926
   Maximum factor change   : 1.317     ["sto1"]
   Maximum relative change: 0.3168     ["sto1"]
                                    .
                                    .
OPTIMISATION ITERATION NO.        :   10
 Model calls so far               :   88
 Starting phi for this iteration: 4.34126E-03
 Contribution to phi from observation group "ogn1": 2.55405E-03
 Contribution to phi from observation group "ogn2": 1.58894E-03
 Contribution to phi from observation group "ogn3": 2.19018E-05
 Contribution to phi from observation group "ogn4": 1.59118E-04
 Contribution to phi from observation group "ogn5": 1.72531E-05
```

```
  All frozen parameters freed

      Lambda =  3.90625E-02 ----->
         Phi =  6.06156E-03  (  1.396 times starting phi)

      Lambda =  1.95313E-02 ----->
         Phi =  6.06208E-03  (  1.396 times starting phi)

      Lambda =  7.81250E-02 ----->
         Phi =  6.06325E-03  (  1.397 times starting phi)

  No more lambdas: phi rising
  Lowest phi this iteration:  6.06156E-03

 Current parameter values          Previous parameter values
   hhc1    1.006178E-04              hhc1     1.009451E-04
   vhc1    5.248568E-05              vhc1     5.131455E-05
   sto1    2.695515E-05              sto1     4.000000E-05
   spy1    0.202028                  spy1     0.199318
  Maximum factor change:    1.484     ["sto1"]
  Maximum relative change:  0.3261    ["sto1"]

OPTIMISATION ITERATION NO.        :  11
 Model calls so far               :  99
 Starting phi for this iteration:  6.06156E-03
 Contribution to phi from observation group "ogn1": 5.66350E-03
 Contribution to phi from observation group "ogn2": 2.93847E-04
 Contribution to phi from observation group "ogn3": 1.28877E-05
 Contribution to phi from observation group "ogn4": 8.18316E-05
 Contribution to phi from observation group "ogn5": 9.49053E-06

      Lambda =  3.90625E-02 ----->
         Phi =  4.87679E-03  (  0.805 of starting phi)

      Lambda =  1.95313E-02 ----->
         Phi =  4.87938E-03  (  0.805 of starting phi)

      Lambda =  7.81250E-02 ----->
         Phi =  4.88598E-03  (  0.806 of starting phi)

 No more lambdas: phi rising
 Lowest phi this iteration:  4.87679E-03
```

```
  Current parameter values          Previous parameter values
    hhc1   1.007928E-04               hhc1     1.006178E-04
    vhc1   5.202901E-05               vhc1     5.248568E-05
    sto1   3.228312E-05               sto1     2.695515E-05
    spy1   0.200343                   spy1     0.202028
   Maximum factor change:   1.198     ["sto1"]
   Maximum relative change: 0.1977    ["sto1"]

 Optimisation complete:  4 optimisation iterations have elapsed
                         since lowest phi was achieved.
   Total model calls  :  110

                          OPTIMISATION RESULTS

Parameters ----->

Parameter       Estimated        95% percent confidence limits
                value            lower limit        upper limit
 hhc1          1.009564E-04      1.005857E-04       1.013285E-04
 vhc1          5.140965E-05      5.020239E-05       5.264595E-05
 sto1          3.981386E-05      3.474760E-05       4.561877E-05
 spy1          0.198272          0.190934           0.205892

Note: confidence limits provide only an indication of parameter
uncertainty. They rely on a linearity assumption which  may not
extend as far in parameter space as the confidence limits
themselves - see PEST manual.

See file C:\PEST\DATA\ATXE18F.SEN for parameter sensitivities.

Observations ----->

Observation    Measured          Calculated         Residual          Weight   Group
                 value             value
 pwc1          0.510000          0.517200          -7.200000E-03      1.000    ogn1
 pwc2          0.970000          0.981100          -1.110000E-02      1.000    ogn1
                                         .
                                         .
 pwc13         2.98000           2.97800            2.000000E-03      1.000    ogn1
 pwc14         3.09000           3.08900            1.000000E-03      1.000    ogn1
 owc11         9.000000E-02      7.004000E-02       1.996000E-02      1.000    ogn2
 owc12         0.204000          0.178200           2.580000E-02      1.000    ogn2
                                         .
                                         .
 owc113        0.942000          0.948000          -6.000000E-03      1.000    ogn2
 owc114        1.05200           1.05800           -6.000000E-03      1.000    ogn2
 owc21         0.00000           0.00000            0.00000           1.000    ogn3
 owc22         5.000000E-04      0.00000            5.000000E-04      1.000    ogn3
                                         .
                                         .
```

```
owc213      9.830000E-02   9.945000E-02  -1.150000E-03   1.000   ogn3
owc214      0.172000       0.175200      -3.200000E-03   1.000   ogn3
owc31       7.000000E-03   3.630000E-03   3.370000E-03   1.000   ogn4
owc32       1.900000E-02   1.329000E-02   5.710000E-03   1.000   ogn4
                                .
                                .
owc313      0.487000       0.488400      -1.400000E-03   1.000   ogn4
owc314      0.612000       0.611900       1.000000E-04   1.000   ogn4
owc41       0.00000        0.00000        0.00000        1.000   ogn5
owc42       1.000000E-04   0.00000        1.000000E-04   1.000   ogn5
owc43       5.000000E-04   0.00000        5.000000E-04   1.000   ogn5
                                .
                                .
owc413      8.400000E-02   8.603000E-02  -2.030000E-03   1.000   ogn5
owc414      0.162000       0.166100      -4.100000E-03   1.000   ogn5

See file C:\PEST\DATA\ATXE18F.RES for more details of residuals
 in graph-ready format.

See file C:\PEST\DATA\ATXE18F.SEO for composite observation
sensitivities.

Objective function ----->

Sum of squared weighted residuals (ie phi)              =  4.2891E-03
Contribution to phi from observation group "ogn1" =  2.5231E-03
Contribution to phi from observation group "ogn2" =  1.5639E-03
Contribution to phi from observation group "ogn3" =  2.6051E-05
Contribution to phi from observation group "ogn4" =  1.5296E-04
Contribution to phi from observation group "ogn5" =  2.3094E-05

Correlation Coefficient ----->

  Correlation coefficient                               =   1.000

Analysis of residuals ----->

 All residuals:-
  Number of residuals with non-zero weight              = 70
  Mean value of non-zero weighted residuals             =  5.0344E-04
  Maximum weighted residual [observation "owc12"]       =  2.5800E-02
  Minimum weighted residual [observation "pwc4"]        = -3.3000E-02
  Standard variance of weighted residuals               =  6.4986E-05
  Standard error of weighted residuals                  =  8.0614E-03

Note: the above variance was obtained by dividing the objective
function by the number of system degrees of freedom (ie. number
of observations with non-zero weight plus number of prior
information articles with non-zero weight minus the number of
```

```
adjustable parameters.) If the degrees of freedom is negative the
divisor becomes the number of observations with non-zero weight
plus the number of prior information items with non-zero weight.

 Residuals for observation group "ogn1":-
  Number of residuals with non-zero weight         = 14
  Mean value of non-zero weighted residuals        = -2.7357E-03
  Maximum weighted residual [observation "pwc9"]   =  1.3000E-02
  Minimum weighted residual [observation "pwc4"]   = -3.3000E-02
  "Variance" of weighted residuals                 =  1.8022E-04
  "Standard error" of weighted residuals           =  1.3425E-02

Note : the above "variance" was obtained by dividing the sum of
squared residuals by the number of items with non-zero weight.

 Residuals for observation group "ogn2":-
  Number of residuals with non-zero weight         = 14
  Mean value of non-zero weighted residuals        =  2.6757E-03
  Maximum weighted residual [observation "owc12"]  =  2.5800E-02
  Minimum weighted residual [observation "owc113"]= -6.0000E-03
  "Variance" of weighted residuals                 =  1.1171E-04
  "Standard error" of weighted residuals           =  1.0569E-02

Note: the above "variance" was obtained by dividing the sum of
squared residuals by the number of items with non-zero weight.

 Residuals for observation group "ogn3":-
  Number of residuals with non-zero weight         = 14
  Mean value of non-zero weighted residuals        =  3.4793E-04
  Maximum weighted residual [observation "owc23"]  =  2.8000E-03
  Minimum weighted residual [observation "owc214"]= -3.2000E-03
  "Variance" of weighted residuals                 =  1.8608E-06
  "Standard error" of weighted residuals           =  1.3641E-03

Note: the above "variance" was obtained by dividing the sum of
squared residuals by the number of items with non-zero weight.

 Residuals for observation group "ogn4":-
  Number of residuals with non-zero weight         = 14
  Mean value of non-zero weighted residuals        =  2.4250E-03
  Maximum weighted residual [observation "owc33"]  =  6.0000E-03
  Minimum weighted residual [observation "owc312"]= -1.8000E-03
  "Variance" of weighted residuals                 =  1.0926E-05
  "Standard error" of weighted residuals           =  3.3054E-03

Note: the above "variance" was obtained by dividing the sum of
squared residuals by the number of items with non-zero weight.
```

```
 Residuals for observation group "ogn5":-
  Number of residuals with non-zero weight            = 14
  Mean value of non-zero weighted residuals           = -1.9571E-04
  Maximum weighted residual [observation "owc44"] =  1.2000E-03
  Minimum weighted residual [observation "owc414"]= -4.1000E-03
  "Variance" of weighted residuals                    =  1.6496E-06
  "Standard error" of weighted residuals              =  1.2844E-03

Note: the above "variance" was obtained by dividing the sum of
squared residuals by the number of items with non-zero weight.

Parameter covariance matrix ----->

             hhc1           vhc1           sto1           spy1
hhc1       6.3955E-07    -3.1422E-06     2.2813E-07    -3.7861E-06
vhc1      -3.1422E-06     2.6680E-05    -1.9552E-05     1.2378E-05
sto1       2.2813E-07    -1.9552E-05     8.7523E-04     4.1488E-06
spy1      -3.7861E-06     1.2378E-05     4.1488E-06     6.7195E-05

Parameter correlation coefficient matrix ----->

             hhc1           vhc1           sto1           spy1
hhc1       1.000         -0.7607         9.6424E-03    -0.5775
vhc1      -0.7607         1.000         -0.1279         0.2923
sto1       9.6424E-03    -0.1279         1.000          1.7108E-02
spy1      -0.5775         0.2923         1.7108E-02     1.000

Normalized eigenvectors of parameter covariance matrix ----->

          Vector_1        Vector_2      Vector_3        Vector_4
hhc1        0.9941         8.7640E-02   -6.3847E-02     3.2238E-04
vhc1        0.1018        -0.9566        0.2722        -2.2954E-02
sto1        1.8393E-03    -2.3315E-02    1.6817E-03     0.9997
spy1        3.7240E-02     0.2770        0.9601         4.7772E-03

Eigenvalues ----->
```

SAMPLE DATA FILE SET 3

Sample Data File Set 3 illustrates MODFLOW-96 and PEST utility program MOD2OBS data file sets generated during the calculation of pumping test drawdown values for an unconfined aquifer simulated as a multilayer aquifer with pumped wellbore storage. MODFLOW-96 and MOD2OBS input and output data file contents are displayed after the conceptual model is briefly described.

The conceptual model pumping test facilities consist of a 100 ft thick unconfined aquifer with a horizontal hydraulic conductivity of 100 ft/day, a vertical hydraulic conductivity of 10 ft/day, a specific yield of 0.2, a storativity of 0.0005, and a specific storage of 0.000005 (see Reilly and Harbaugh, 1993a). The MODFLOW-96 areal grid has 11 layers, 50 rows, and 50 columns. The layer thickness, designed so that delayed drainage can be simulated, is 5.0 ft in Layer 1 and 11 and 10 feet in the rest of the layers. The first layer is assigned a specific yield of 0.2 and the other layers are assigned a storativity based on a specific storage of 0.000005 and the layer thickness.

The multiple layer well is partially penetrating and is screened in the bottom 25 ft of the aquifer. Pumped wellbore storage is appreciable. The well effective radius is 0.936 ft. The constant well discharge is 125,670 ft^3/day. The nodes in Layer 11, Layer 10, and Layer 9 at Row 25 and Column 25 representing the well are defined to have a discharge of 25,130, 50,270, and 50,270 ft^3/day, respectively. Thus, the multiple layer pumped well is simulated as three pumped wells at the same location. Measured drawdowns for a point 55 ft deep at a radius of 16 ft (Row 25, between Column 25 and Column 26) are of interest. The real-world starting date of the pumping test is 06/15/1994 and the starting time is 08:00:00. The east and north coordinates of the top left corner of the MODFLOW-96 grid are 200,000 and 200,000 ft, respectively. The east and north coordinates of the observation well are 164,616 and 35,400 ft, respectively. The rotation of the MODFLOW-96 grid row direction is 0.0°.

MODFLOW-96 BCF INPUT DATA FILE

The areal grid has the variable row and column grid spacings as indicated in the following MODFLOW-96 BCF input data file contents:

```
         0          0
 0 0 0 0 0 0 0 0 0 0 0
         0        1.0
INTERNAL 1.0 (FREE) 0
  21000.  14000.    9400.   6300.   4200.   2800.   1900.   1300.    900.
    650.    450.    300.    200.    150.    100.    100.    100.    100.
    100.    100.    100.    100.    100.    100.    100.    100.    100.
    100.    100.    100.    100.    100.    100.    100.    100.    100.
    150.    200.    300.    450.    650.    900.   1300.   1900.   2800.
   4200.   6300.   9400.  14000.  21000.
INTERNAL 1.0 (FREE) 0
  21000.  14000.    9400.   6300.   4200.   2800.   1900.   1300.    900.
    650.    450.    300.    200.    150.    100.    100.    100.    100.
    100.    100.    100.    100.    100.    100.    100.    100.    100.
    100.    100.    100.    100.    100.    100.    100.    100.    100.
    150.    200.    300.    450.    650.    900.   1300.   1900.   2800.
   4200.   6300.   9400.  14000.  21000.
         0         .2
         0     500.00
         0       1.33
         0     .00005
         0    1000.00
         0       1.00
         0     .00005
         0    1000.00
         0       1.00
```

```
         0    .00005
         0   1000.00
         0      1.00
         0    .00005
         0   1000.00
         0      1.00
         0    .00005
         0   1000.00
         0      1.00
         0    .00005
         0   1000.00
         0      1.00
         0    .00005
         0   1000.00
         0      1.00
         0    .00005
         0   1000.00
         0      1.00
         0    .00005
         0   1000.00
         0      1.00
         0    .00005
         0    500.00
```

MODFLOW-96 BASIC INPUT DATA FILE

Time is subdivided into 45 logarithmically spaced stress periods with the elapsed times as indicated in the following MODFLOW-96 basic input data file contents:

```
Sample Data File Set 3
        11         50         50         45          4
FREE
         0          1
         0          1
         0          1
         0          1
         0          1
         0          1
         0          1
         0          1
         0          1
         0          1
         0          1
         0          1
    999.99
         0       700.
         0       700.
         0       700.
         0       700.
         0       700.
         0       700.
         0       700.
         0       700.
         0       700.
         0       700.
         0       700.
   0.00002          1        1.0
   0.00002          1        1.0
   0.00002          1        1.0
```

```
0.00002        1       1.0
0.00002        1       1.0
0.00010        1       1.0
0.00010        1       1.0
0.00010        1       1.0
0.00010        1       1.0
0.00020        1       1.0
0.00020        1       1.0
0.00020        1       1.0
0.00020        1       1.0
0.00020        1       1.0
0.00050        1       1.0
0.00050        1       1.0
0.00050        1       1.0
0.00200        1       1.0
0.00200        1       1.0
0.00200        1       1.0
0.00200        1       1.0
0.00200        1       1.0
0.00200        1       1.0
0.00500        1       1.0
0.00500        1       1.0
0.00500        1       1.0
0.00500        1       1.0
0.00500        1       1.0
0.01000        1       1.0
0.01000        1       1.0
0.01000        1       1.0
0.01000        1       1.0
0.01000        1       1.0
0.01000        1       1.0
0.05000        1       1.0
0.05000        1       1.0
0.05000        1       1.0
0.05000        1       1.0
0.10000        1       1.0
0.10000        1       1.0
0.10000        1       1.0
0.10000        1       1.0
0.10000        1       1.0
0.10000        1       1.0
0.10000        1       1.0
```

MODFLOW-96 WELL INPUT DATA FILE

Discharge rates are distributed to layers as indicated in the following MODFLOW-96 well input data file contents:

```
 3        0
 3
 9       25         25     14076.
10       25         25     14076.
11       25         25     7036.
 3
```

```
    9         25         25     29157.
   10         25         25     29157.
   11         25         25     14575.
                    .
                    .
    3
    9         25         25     50270.
   10         25         25     50270.
   11         25         25     25130.
    3
    9         25         25     50270.
   10         25         25     50270.
   11         25         25     25130.
```

MODFLOW-96 STRONGLY IMPLICIT PROCEDURE INPUT DATA FILE

```
   50          5
  1.0       .001         0       .001         1
```

MODFLOW-96 OUTPUT CONTROL INPUT DATA FILE

```
DRAWDOWN PRINT FORMAT  0
DRAWDOWN SAVE FORMAT
DRAWDOWN SAVE UNIT  61
COMPACT BUDGET FILES
PERIOD 1 STEP 1
SAVE DRAWDOWN 6
PERIOD 2 STEP 1
SAVE DRAWDOWN 6
       .
       .
PERIOD 44 STEP 1
SAVE DRAWDOWN 6
PERIODd 45 STEP 1
PRINT DRAWDOWN 6
SAVE DRAWDOWN 6  1
```

MODFLOW-96 NAME INPUT DATA FILE

```
LIST            6   c:\modflow96\data\samp3lst.txt
BAS             5   c:\modflow96\data\samp3bas.txt
OC             24   c:\modflow96\data\samp3oc.txt
```

```
BCF             21   c:\modflow96\data\samp3bcf.txt
WEL             22   c:\modflow96\data\samp3wel.txt
SIP             23   c:\modflow96\data\samp3sip.txt
DATA(BINARY)    61   c:\modflow96\data\samp3dd.bin
```

MOD2OBS MEASURED BORE SAMPLE INPUT DATA FILE

```
ow1  06/15/1994  08:00:02  -0.01
ow1  06/15/1994  08:00:03  -0.04
                .
                .
ow1  06/16/1994  05:36:00  -3.64
ow1  06/16/1994  08:00:00  -3.09
```

MOD2OBS GRID SPECIFICATION FILE

```
50 50
200000. 200000. 0.0
21000  14000  9400  6300  4200  2800  1900  1300  900
650  450  300  200  150  100  100  100  100
100  100  100  100  100  100  100  100  100
100  100  100  100  100  100  100  100  100
150  200  300  450  650  900  1300  1900  2800
4200  6300  9400  14000  21000
21000  14000  9400  6300  4200  2800  1900  1300  900
650  450  300  200  150  100  100  100  100
100  100  100  100  100  100  100  100  100
100  100  100  100  100  100  100  100  100
150  200  300  450  650  900  1300  1900  2800
4200  6300  9400  14000  21000
```

MOD2OBS BORE COORDINATE FILE

```
ow1  164616    35400    6
```

MOD2OBS OUTPUT FILE

```
OW1        06/15/1994    08:00:02    -1.6539754E-02
OW1        06/15/1994    08:00:03    -3.3537328E-02
                              .
                              .
OW1        06/16/1994    05:36:00    -3.744365
OW1        06/16/1994    08:00:00    -3.801044
```

SMP2HYD OUTPUT FILE

```
TIME_IN_DAYS      DATE          TIME          BORE_OW1
2.314815E-05   06/15/1994    08:00:02     -1.653975E-02
3.472222E-05   06/15/1994    08:00:03     -3.353733E-02
                                 .

                                 .
0.900000       06/16/1994    05:36:00     -3.74437
1.00000        06/16/1994    08:00:00     -3.80104
```

SAMPLE DATA FILE SET 4

Sample Data File Set 4 illustrates MODFLOW-96 data sets generated during the calculation of cylindrical flow pumping test drawdown values for an unconfined aquifer with 10 confined layers and 1 unconfined layer. Time-drawdown values with partially penetrating well conditions are calculated. MODFLOW-96, MOD2OBS, and SMP2HYD input and output data file contents are displayed after the conceptual model is briefly described.

The conceptual modeling areal flow pumping test facilities consist of a 100 ft thick unconfined aquifer. Parameter values are horizontal hydraulic conductivity = 100 ft/day, vertical hydraulic conductivity = 10 ft/day, specific yield = 0.2, storativity = 0.0005, and specific storage = 0.000005 1/ft (see Reilly and Harbaugh, 1993a). The MODFLOW-96 grid has 1 layer, 11 rows, and 40 columns radially spaced with a multiplier of 1.5. The multilayer pumped well is partially penetrating and is screened in the bottom 25 ft of the aquifer. Pumped wellbore storage is negligible. The pumped well effective radius is 0.936 ft. The constant pumped well discharge is 125,670 ft^3/day.

The length of the pumping test is 1 day. Time is subdivided into 45 logarithmically spaced stress periods. The nodes in Layer 11, Layer 10, and Layer 9 of Column 1 representing the multilayer pumped well are defined to have a discharge of 25,130, 50,270, and 50,270 ft^3/day, respectively. The multilayer pumped well is simulated as three wells at the same node. A constant zero drawdown is simulated at the far radial boundary (Column 40), which is at a distance of 6.9E6 ft. Measured drawdowns are for a point 50 ft deep at a radius of 16 ft (Column 8 and Row 6; see Reilly and Harbaugh, 1993a, p. 493). Time is subdivided into 45 logarithmically spaced stress periods.

MODFLOW-96 NAME FILE

```
LIST            6    c:\modflow96\data\samp4lst.txt
BAS             5    c:\modflow96\data\samp4bas.txt
OC             24    c:\modflow96\data\samp4oc.txt
WEL            22    c:\modflow96\data\samp4wel.txt
GFD            21    c:\modflow96\data\samp4gfd.txt
SIP            23    c:\modflow96\data\samp4sip.txt
DATA(BINARY)   61    c:\modflow96\data\samp4dd.bin
```

MODFLOW-96 BASIC FILE

```
Sample Data File Set 4
Cylindrical Well
        1         11         40         45          4

        0          1
        5          1                (25I3)          0
1  1  1  1  1  1  1  1  1  1  1  1  1  1  1  1  1  1  1  1  1  1  1  1  1
1  1  1  1  1  1  1  1  1  1  1  1  1  1 -1
1  1  1  1  1  1  1  1  1  1  1  1  1  1  1  1  1  1  1  1  1  1  1  1  1
                                         .

                                         .
1  1  1  1  1  1  1  1  1  1  1  1  1  1  1  1  1  1  1  1  1  1  1  1  1
1  1  1  1  1  1  1  1  1  1  1  1  1  1 -1
1  1  1  1  1  1  1  1  1  1  1  1  1  1  1  1  1  1  1  1  1  1  1  1  1
1  1  1  1  1  1  1  1  1  1  1  1  1  1 -1
    999.99
         0      700.
   0.00002         1       1.0
   0.00002         1       1.0
   0.00002         1       1.0
   0.00002         1       1.0
   0.00002         1       1.0
   0.00010         1       1.0
                 .

                 .
   0.10000         1       1.0
   0.10000         1       1.0
   0.10000         1       1.0
```

MODFLOW-96 GENERAL FINITE-DIFFERENCE FLOW FILE

```
        0         0
 0
        21        1.0            (6E12.4)         0
  1.0586E+00  4.7637E+00  1.0718E+01  2.4116E+01  5.4261E+01  1.2209E+02
  2.7470E+02  6.1807E+02  1.3907E+03  3.1290E+03  7.0402E+03  1.5840E+04
  3.5641E+04  8.0192E+04  1.8043E+05  4.0597E+05  9.1344E+05  2.0552E+06
  4.6243E+06  1.0405E+07  2.3410E+07  5.2673E+07  1.1852E+08  2.6666E+08
  5.9998E+08  1.3500E+09  3.0374E+09  6.8342E+09  1.5377E+10  3.4598E+10
  7.7846E+10  1.7515E+11  3.9409E+11  8.8671E+11  1.9951E+12  4.4890E+12
  1.0100E+13  2.2725E+13  5.1132E+13  5.7524E+13
        21        1.0              (25I3)         0
  1  0  0  0  0  0  0  0  0  0
        21        1.0            (6E12.4)         0
  2.1172E-01  9.5273E-01  2.1437E+00  4.8232E+00  1.0852E+01  2.4418E+01
  5.4939E+01  1.2361E+02  2.7813E+02  6.2579E+02  1.4080E+03  3.1681E+03
  7.1282E+03  1.6038E+04  3.6086E+04  8.1195E+04  1.8269E+05  4.1105E+05
  9.2486E+05  2.0809E+06  4.6821E+06  1.0535E+07  2.3703E+07  5.3332E+07
  1.2000E+08  2.6999E+08  6.0748E+08  1.3668E+09  3.0754E+09  6.9196E+09
  1.5569E+10  3.5031E+10  7.8819E+10  1.7734E+11  3.9902E+11  8.9780E+11
  2.0200E+12  4.5451E+12  1.0226E+13  1.1505E+13
                                                  .

                                                  .
```

```
2.6465E-05  1.1909E-04  2.6796E-04  6.0290E-04  1.3565E-03  3.0522E-03
6.8674E-03  1.5452E-02  3.4766E-02  7.8224E-02  1.7600E-01  3.9601E-01
8.9102E-01  2.0048E+00  4.5108E+00  1.0149E+01  2.2836E+01  5.1381E+01
1.1561E+02  2.6012E+02  5.8526E+02  1.3168E+03  2.9629E+03  6.6665E+03
1.5000E+04  3.3749E+04  7.5935E+04  1.7085E+05  3.8442E+05  8.6495E+05
1.9461E+06  4.3788E+06  9.8524E+06  2.2168E+07  4.9878E+07  1.1222E+08
2.5250E+08  5.6814E+08  1.2783E+09  1.4381E+09
      21        1.0             (7F10.2)         0
 7748.14   7748.14   7748.14   7748.14   7748.14   7748.14   7748.14
 7748.14   7748.14   7748.14   7748.14   7748.14   7748.14   7748.14
 7748.14   7748.14   7748.14   7748.14   7748.14   7748.14   7748.14
 7748.14   7748.14   7748.14   7748.14   7748.14   7748.14   7748.14
 7748.14   7748.14   7748.14   7748.14   7748.14   7748.14   7748.14
 7748.14   7748.14   7748.14   7748.14      0.00
                                  .
                                  .
 7748.14   7748.14   7748.14   7748.14   7748.14   7748.14   7748.14
 7748.14   7748.14   7748.14   7748.14   7748.14   7748.14   7748.14
 7748.14   7748.14   7748.14   7748.14   7748.14   7748.14   7748.14
 7748.14   7748.14   7748.14   7748.14   7748.14   7748.14   7748.14
 7748.14   7748.14   7748.14   7748.14   7748.14   7748.14   7748.14
 7748.14   7748.14   7748.14   7748.14      0.00
      21        1.0             (6E12.4)         0
1.0586E+00  4.7637E+00  1.0718E+01  2.4116E+01  5.4261E+01  1.2209E+02
2.7470E+02  6.1807E+02  1.3907E+03  3.1290E+03  7.0402E+03  1.5840E+04
3.5641E+04  8.0192E+04  1.8043E+05  4.0597E+05  9.1344E+05  2.0552E+06
4.6243E+06  1.0405E+07  2.3410E+07  5.2673E+07  1.1852E+08  2.6666E+08
5.9998E+08  1.3500E+09  3.0374E+09  6.8342E+09  1.5377E+10  3.4598E+10
7.7846E+10  1.7515E+11  3.9409E+11  8.8671E+11  1.9951E+12  4.4890E+12
1.0100E+13  2.2725E+13  5.1132E+13  5.7524E+13
                                  .
                                  .
0.0000E+00  0.0000E+00  0.0000E+00  0.0000E+00  0.0000E+00  0.0000E+00
0.0000E+00  0.0000E+00  0.0000E+00  0.0000E+00  0.0000E+00  0.0000E+00
0.0000E+00  0.0000E+00  0.0000E+00  0.0000E+00  0.0000E+00  0.0000E+00
0.0000E+00  0.0000E+00  0.0000E+00  0.0000E+00  0.0000E+00  0.0000E+00
0.0000E+00  0.0000E+00  0.0000E+00  0.0000E+00  0.0000E+00  0.0000E+00
0.0000E+00  0.0000E+00  0.0000E+00  0.0000E+00  0.0000E+00  0.0000E+00
0.0000E+00  0.0000E+00  0.0000E+00  0.0000E+00
```

MODFLOW-96 WELL FILE

```
3         0
3
1         9          1     50270.
1        10          1     50270.
1        11          1     25130.
3
1         9          1     50270.
1        10          1     50270.
1        11          1     25130.
                 .
                 .
```

```
        3
        1         9         1    50270.
        1        10         1    50270.
        1        11         1    25130.
        3
        1         9         1    50270.
        1        10         1    50270.
        1        11         1    25130.
```

MODFLOW-96 STRONGLY IMPLICIT PROCEDURE FILE

```
       50         5
      1.0      .001         0      .001         1
```

MODFLOW-96 OUTPUT CONTROL FILE

```
DRAWDOWN PRINT FORMAT  0
DRAWDOWN SAVE FORMAT
DRAWDOWN SAVE UNIT  61
COMPACT BUDGET FILES
PERIOD 1 STEP 1
SAVE DRAWDOWN
PERIOD 2 STEP 1
       .
       .
PERIOD 44 STEP 1
SAVE DRAWDOWN
PERIOD 45 STEP 1
PRINT DRAWDOWN
SAVE DRAWDOWN
```

MODFLOW-96 PRIMARY STORAGE CAPACITY FILE

```
2.117E-01 9.527E-01 2.144E+00 4.823E+00 1.085E+01 2.442E+01 5.494E+01
1.236E+02 2.781E+02 6.258E+02 1.408E+03 3.168E+03 7.128E+03 1.604E+04
3.609E+04 8.119E+04 1.827E+05 4.110E+05 9.249E+05 2.081E+06 4.682E+06
1.053E+07 2.370E+07 5.333E+07 1.200E+08 2.700E+08 6.075E+08 1.367E+09
3.075E+09 6.920E+09 1.557E+10 3.503E+10 7.882E+10 1.773E+11 3.990E+11
8.978E+11 2.020E+12 4.545E+12 1.023E+13 1.150E+13
                                     .
                                     .
2.646E-05 1.191E-04 2.680E-04 6.029E-04 1.357E-03 3.052E-03 6.867E-03
1.545E-02 3.477E-02 7.822E-02 1.760E-01 3.960E-01 8.910E-01 2.005E+00
4.511E+00 1.015E+01 2.284E+01 5.138E+01 1.156E+02 2.601E+02 5.853E+02
1.317E+03 2.963E+03 6.666E+03 1.500E+04 3.375E+04 7.594E+04 1.709E+05
3.844E+05 8.650E+05 1.946E+06 4.379E+06 9.852E+06 2.217E+07 4.988E+07
1.122E+08 2.525E+08 5.681E+08 1.278E+09 1.438E+09
```

MODFLOW-96 CONDUCTANCE ALONG ROWS FILE

```
  7748.14    7748.14    7748.14    7748.14    7748.14    7748.14    7748.14
  7748.14    7748.14    7748.14    7748.14    7748.14    7748.14    7748.14
  7748.14    7748.14    7748.14    7748.14    7748.14    7748.14    7748.14
  7748.14    7748.14    7748.14    7748.14    7748.14    7748.14    7748.14
  7748.14    7748.14    7748.14    7748.14    7748.14    7748.14    7748.14
  7748.14    7748.14    7748.14    7748.14       0.00
 15496.28   15496.28   15496.28   15496.28   15496.28   15496.28   15496.28
 15496.28   15496.28   15496.28   15496.28   15496.28   15496.28   15496.28
 15496.28   15496.28   15496.28   15496.28   15496.28   15496.28   15496.28
 15496.28   15496.28   15496.28   15496.28   15496.28   15496.28   15496.28
 15496.28   15496.28   15496.28   15496.28   15496.28   15496.28   15496.28
 15496.28   15496.28   15496.28   15496.28       0.00
                                          .
                                          .
 15496.28   15496.28   15496.28   15496.28   15496.28   15496.28   15496.28
 15496.28   15496.28   15496.28   15496.28   15496.28   15496.28   15496.28
 15496.28   15496.28   15496.28   15496.28   15496.28   15496.28   15496.28
 15496.28   15496.28   15496.28   15496.28   15496.28   15496.28   15496.28
 15496.28   15496.28   15496.28   15496.28   15496.28   15496.28   15496.28
 15496.28   15496.28   15496.28   15496.28       0.00
  7748.14    7748.14    7748.14    7748.14    7748.14    7748.14    7748.14
  7748.14    7748.14    7748.14    7748.14    7748.14    7748.14    7748.14
  7748.14    7748.14    7748.14    7748.14    7748.14    7748.14    7748.14
  7748.14    7748.14    7748.14    7748.14    7748.14    7748.14    7748.14
  7748.14    7748.14    7748.14    7748.14    7748.14    7748.14    7748.14
  7748.14    7748.14    7748.14    7748.14       0.00
```

MODFLOW-96 CONDUCTANCE ALONG COLUMNS FILE

```
1.0586E+00  4.7637E+00  1.0718E+01  2.4116E+01  5.4261E+01  1.2209E+02
2.7470E+02  6.1807E+02  1.3907E+03  3.1290E+03  7.0402E+03  1.5840E+04
3.5641E+04  8.0192E+04  1.8043E+05  4.0597E+05  9.1344E+05  2.0552E+06
4.6243E+06  1.0405E+07  2.3410E+07  5.2673E+07  1.1852E+08  2.6666E+08
5.9998E+08  1.3500E+09  3.0374E+09  6.8342E+09  1.5377E+10  3.4598E+10
7.7846E+10  1.7515E+11  3.9409E+11  8.8671E+11  1.9951E+12  4.4890E+12
1.0100E+13  2.2725E+13  5.1132E+13  5.7524E+13
                                          .
                                          .
7.0573E-01  3.1758E+00  7.1455E+00  1.6077E+01  3.6174E+01  8.1392E+01
1.8313E+02  4.1205E+02  9.2710E+02  2.0860E+03  4.6935E+03  1.0560E+04
2.3761E+04  5.3461E+04  1.2029E+05  2.7065E+05  6.0896E+05  1.3702E+06
3.0829E+06  6.9364E+06  1.5607E+07  3.5116E+07  7.9010E+07  1.7777E+08
3.9999E+08  8.9998E+08  2.0249E+09  4.5561E+09  1.0251E+10  2.3065E+10
5.1897E+10  1.1677E+11  2.6273E+11  5.9114E+11  1.3301E+12  2.9927E+12
6.7335E+12  1.5150E+13  3.4088E+13  3.8349E+13
0.0000E+00  0.0000E+00  0.0000E+00  0.0000E+00  0.0000E+00  0.0000E+00
0.0000E+00  0.0000E+00  0.0000E+00  0.0000E+00  0.0000E+00  0.0000E+00
0.0000E+00  0.0000E+00  0.0000E+00  0.0000E+00  0.0000E+00  0.0000E+00
0.0000E+00  0.0000E+00  0.0000E+00  0.0000E+00  0.0000E+00  0.0000E+00
0.0000E+00  0.0000E+00  0.0000E+00  0.0000E+00  0.0000E+00  0.0000E+00
0.0000E+00  0.0000E+00  0.0000E+00  0.0000E+00  0.0000E+00  0.0000E+00
0.0000E+00  0.0000E+00  0.0000E+00  0.0000E+00
```

MODFLOW-96 LATERAL DISTANCE FROM PUMPED WELL FILE

```
9.3600E-01  1.4040E+00  2.1060E+00  3.1590E+00  4.7385E+00  7.1077E+00
1.0662E+01  1.5992E+01  2.3989E+01  3.5983E+01  5.3974E+01  8.0962E+01
1.2144E+02  1.8216E+02  2.7325E+02  4.0987E+02  6.1480E+02  9.2220E+02
1.3833E+03  2.0750E+03  3.1124E+03  4.6687E+03  7.0030E+03  1.0504E+04
1.5757E+04  2.3635E+04  3.5453E+04  5.3179E+04  7.9768E+04  1.1965E+05
1.7948E+05  2.6922E+05  4.0383E+05  6.0574E+05  9.0861E+05  1.3629E+06
2.0444E+06  3.0666E+06  4.5998E+06  6.8998E+06
```

MODFLOW-96 LAYER THICKNESS FILE

```
5.0000E+00  1.0000E+01  1.0000E+01  1.0000E+01  1.0000E+01  1.0000E+01
1.0000E+01  1.0000E+01  1.0000E+01  1.0000E+01  5.0000E+00
```

MOD2OBS GRID SPECIFICATION FILE

```
11 40
100000. 1000. 0.0
1.0586 4.7637  10.718  24.116  54.261  122.09
274.7 618.07  1390.7  3129  7040.2  15840
35641 80192  180430  405970  913440  2055200
4624300 1.0405E+07  2.341E+07  5.2673E+07  1.1852E+08  2.6666E+08
5.9998E+08 1.35E+09  3.0374E+09  6.8342E+09  1.5377E+10  3.4598E+10
7.7846E+10 1.7515E+11  3.9409E+11  8.8671E+11  1.9951E+12  4.489E+12
1.01E+13 2.2725E+13  5.1132E+13  5.7524E+13
5. 10. 10. 10. 10. 10. 10. 10. 10. 10. 5.
```

MOD2OBS BORE COORDINATE FILE

```
ow1  100016.    950.    1
```

MOD2OBS BORE SAMPLE TIME-DRAWDOWN FILE

```
OW1          06/15/1996    08:00:13     -0.4330888
OW1          06/15/1996    08:00:33     -1.106119
                         .
                         .
OW1          06/15/1996    23:21:36     -3.318587
OW1          06/16/1996    08:00:00     -3.526169
```

MOD2OBS OUTPUT FILE

```
OW1          06/15/1996    08:00:13     -0.4330888
OW1          06/15/1996    08:00:33     -1.106119
                         .
                         .
OW1          06/15/1996    23:21:36     -3.318587
OW1          06/16/1996    08:00:00     -3.526169
```

SMP2HYD OUTPUT FILE

```
TIME_IN_DAYS      DATE          TIME        BORE_OW1
1.504630E-04   06/15/1996    08:00:13     -0.433089
3.819444E-04   06/15/1996    08:00:33     -1.10612
                                 .
                                 .
0.640000       06/15/1996    23:21:36     -3.31859
1.00000        06/16/1996    08:00:00     -3.52617
```

SAMPLE DATA FILE SET 5

Sample Data File Set 5 illustrates PEST data sets associated with inferring heterogeneity using MODFLOW-96 and PEST regularization and pilot point capabilities. Only PEST pilot point, control, PPK2FAC, and FAC2REAL file contents are displayed after the conceptual model is briefly described because MODFLOW-96 data file sets have been covered in previous sample data file sets.

Briefly, the conceptual modeling pumping test facilities consist of a 100 ft thick unconfined aquifer with a horizontal hydraulic conductivity of 100 ft/day except in a narrow ellipse shaped area surrounding the pumped well where it is 200 ft/day. The aquifer has a uniform specific yield of 0.2. The elliptical area of high hydraulic conductivity averages about 300 ft wide and trends at an angle of 45° from north through the pumped well. The MODFLOW-96 areal grid has 1 layer, 50 rows, and 50 columns. Delayed drainage is not simulated.

The pumped well is partially penetrating and is screened in the bottom 25 ft of the aquifer. Pumped wellbore storage is not simulated. The well effective radius is 0.936 ft. The constant well discharge is 125,670 ft^3/day. Time is subdivided into 45 logarithmically spaced stress periods each with 1 time step. Drawdown data for nine observation wells scattered in and around the high conductivity area within 1000 ft of the pumped well are available. The observation wells are 55 ft deep and have short screens.

The real-world starting date of the pumping test is 06/15/1994 and the starting time is 08:00:00. The east and north coordinates of the top left corner of the MODFLOW-96 grid are 200,000 and 200,000 ft, respectively. The rotation of the MODFLOW-96 grid row direction is 0.0°. The east and north coordinates of the observation wells are given in a PEST bore coordinate file. The bore coordinate file is not illustrated here because bore coordinate files were covered in previous sample data file sets. The rotation of the MODFLOW-96 grid row direction is 0.0°.

There are 31 pilot points scattered around the pumped well. Most of the pilot points are within 1000 ft of the pumped well. Pilot point coordinates are given in the pilot point file.

COMPOSITE MODEL FILE

```
c:\pest\Utility\ppk2fac.exe < c:\pest\utility\pt2ppk2.txt
c:\pest\utility\fac2real.exe < c:\pest\utility\pt2f2r.txt
c:\Modflow96\mf96hyd.exe < c:\pest\utility\pt2nam.txt
c:\pest\Utility\mod2obs3.exe < c:\pest\utility\pt2m2rpp.txt
```

PILOT POINT FILE

```
pp1 264700. 135400. 4 15651.748
pp2 265010. 135400. 3 12000.000
                .
                .
pp30 264200. 135900. 3 11088.481
pp31 264600. 135400. 4 15932.118
```

PILOT POINT TEMPLATE FILE

```
ptf #
pp1 264700. 135400. 4 #tra1   #
pp2 265010. 135400. 3 #tra2   #
                .
                .
pp30 264200. 135900. 3 #tra30  #
pp31 264600. 135400. 4 #tra31  #
```

PILOT POINT INSTRUCTION FILE

```
pif @
l1 w w w w !tra1!
l1 w w w w !tra2!
         .
         .
l1 w w w w !tra30!
l1 w w w w !tra31!
```

PEST CONTROL FILE

```
pcf
* control data
restart regularisation
31 405 1 31 10
1 1 single point 1 0 0
10.0 2.0 0.3 0.03 10
```

```
10.0 10.0 0.001
0.1
30 0.01 3 3 0.01 3
1 1 1
* parameter groups
pgn1 relative .01 0.0 switch 1.0 parabolic
* parameter data
tra1 log factor 19152. 15000. 25000. pgn1 1 0 1
tra2 log factor  8928. 5000. 12000. pgn1 1 0 1
                         .
                         .
tra30 log factor  8928. 5000. 12000. pgn1 1 0 1
tra31 log factor 19152. 15000. 25000. pgn1 1 0 1
* observation groups
ogn1
ogn2
ogn3
ogn4
ogn5
ogn6
ogn7
ogn8
ogn9
regul
* observation data
OBS11  -1.0524006E-05  1.0  ogn1
OBS12  -1.0666009E-05  1.0  ogn1
                .
                .
OBS144  -1.826102  1.0  ogn1
OBS145  -1.904076  1.0  ogn1
OBS21  -1.0240000E-05  1.0  ogn2
OBS22  -1.0240000E-05  1.0  ogn2
                .
                .
OBS244  -0.3136270  1.0  ogn2
OBS245  -0.3561519  1.0  ogn2
OBS31  -1.0481176E-05  1.0  ogn3
OBS32  -1.0601765E-05  1.0  ogn3
                .
                .
OBS344  -1.756577  1.0  ogn3
OBS345  -1.834370  1.0  ogn3
OBS41  -1.0240000E-05  1.0  ogn4
OBS42  -1.0240000E-05  1.0  ogn4
                .
                .
```

```
OBS444  -0.5400326  1.0  ogn4
OBS445 -0.5961924  1.0  ogn4
OBS51  -1.0524011E-05  1.0  ogn5
OBS52  -1.0666016E-05  1.0  ogn5
                .
                .
OBS544  -1.839737  1.0  ogn5
OBS545  -1.918088  1.0  ogn5
OBS61  -1.0240000E-05  1.0  ogn6
OBS62  -1.0240000E-05  1.0  ogn6
                .
                .
OBS644  -0.6096196  1.0  ogn6
OBS645  -0.6694604  1.0  ogn6
OBS71  -1.0524010E-05  1.0  ogn7
OBS72  -1.0666015E-05  1.0  ogn7
                .
                .
OBS744  -1.837318  1.0  ogn7
OBS745  -1.915573  1.0  ogn7
OBS81  -1.0240000E-05  1.0  ogn8
OBS82  -1.0240000E-05  1.0  ogn8
                .
                .
OBS844  -0.6068601  1.0  ogn8
OBS845  -0.6662883  1.0  ogn8
OBS91  -1.2040108E-03  1.0  ogn9
OBS92  -1.8008963E-03  1.0  ogn9
                .
                .
OBS944  -3.152408  1.0  ogn9
OBS945  -3.232368  1.0  ogn9
* model command line
c:\pest\utility\pt2ppr.bat
* model input/output
c:\pest\utility\pt2temppil.txt c:\pest\utility\pt2pp.txt
c:\pest\utility\pt2inspp.txt c:\pest\utility\pt2m2oppr.txt
* prior information
pi1 1.0 * log(tra1) - 1.0 * log(tra2) = 0.0  1.0  regul
pi2 1.0 * log(tra2) - 1.0 * log(tra3) = 0.0  1.0  regul
                               .
                               .
pi30 1.0 * log(tra30) - 1.0 * log(tra31) = 0.0  1.0  regul
pi31 1.0 * log(tra31) - 1.0 * log(tra1) = 0.0  1.0  regul
* regularisation
0.15 0.17  0.0
1.0E-2  1.0E-6  1.0E6
1.3  1.0E-2
```

PEST KRIGING STRUCTURE FILE

```
STRUCTURE str1
NUGGET 0.0
TRANSFORM NONE
NUMVARIOGRAM 1
VARIOGRAM var1 1.0
END STRUCTURE
VARIOGRAM var1
VARTYPE 2
BEARING 45.0
ANISOTROPY 1.0
A 500
END VARIOGRAM
```

PEST PPK2FAC REDIRECT FILE

```
c:\pest\utility\pt2gs.txt
c:\pest\utility\pt2pp.txt
0.0
c:\pest\utility\pt2int.txt
c:\pest\utility\pt2ks.txt
str1
o
10000.
1
31
str1
o
1000.
1
31
c:\pest\utility\pt2ifact.txt
f
c:\pest\utility\pt2sd.ref
c:\pest\utility\pt2outreg.txt
```

PEST FAC2REAL REDIRECT FILE

```
c:\pest\utility\pt2ifact.txt
f
c:\pest\utility\pt2pp.txt
s
8000.
s
25000.
c:\pest\utility\pt2zarray.txt
f
1035
```

References

Abramowitz, M. and I.A. Stegun, Eds., 1964, *Handbook of Mathematical Functions with Formulas, Graphs, and Mathematical Tables.* Applied Mathematical Series, Vol. 55, U.S. Department of Commerce, National Bureau of Standards, Washington, D.C., 1046 pp.

Akindunni, F.F. and R.W. Gillham, 1992, Unsaturated and saturated flow in response to pumping of an unconfined aquifer: numerical investigation of delayed drainage, *Ground Water,* 30, pp. 873–884.

Andersen, P.F., 1993, A Manual of Instructional Problems for the U.S.G.S. MODFLOW Model, U.S. Environmental Agency, Report EPA/600/R-93/010.

Anderson, M.P. and W.W. Woessner, 1992, *Applied Ground Water Modeling: Simulation of Flow and Advective Transport,* Academic Press, Inc., New York.

Barlow, P.M. and A.F. Moench, 1999, WTAQ: A Computer Program for Calculating Drawdowns and Estimating Hydraulic Properties for Confined and Water-Table Aquifers, U.S. Geological Survey Water-Resources Investigations Report 99-4225, 74 pp.

Batu, V., 1998, *Aquifer Hydraulics: A Comprehensive Guide to Hydrological Data Analysis*, John Wiley Interscience Publications, Somerset, NJ.

Bear, J., 1972, *Dynamics of Fluids in Porous Media*, American Elsevier, New York, 764 pp.

Bear, J. and A. Verruijt, 1987, *Modeling Ground Water Flow and Pollution,* D. Reidel Publishing Co., New York, 408 pp.

Beljin, M.S. 1987, Representation of Individual Wells in Two-Dimensional Ground Water Modeling, in *Proceedings of the NWWA Conference on Solving Ground Water Problems with Models*, National Water Well Association, Westerville, OH, pp. 340–351.

Bierschenk, W.H., 1963, Determining Well Efficiency by Multiple Step-Drawdown Tests, Publication 64, International Association of Scientific Hydrology, pp. 493–507.

Boak, R.A., 1991, Auger Hole, Piezometer, and Slug Test: A Literature Review, M.Sc. thesis, University of Newcastle upon Tyne, UK.

Bohling, G.C., X. Zhan, J.J. Butler, and L. Zheng, 2002, Steady-shape analysis of tomographic pumping tests for characterization of aquifer heterogeneities, *Water Resour. Res.*, 38(12), 1324; doi: 10.1029/2001WRR001176.

Bohling, G.C., X. Zhan, M.D. Knoll, and J.J. Butler, 2003, Hydraulic Tomography and the Impact of *A Priori* Information: an Alluvial Aquifer Example, Kansas State Geological Survey Open-File Report 2003-71.

Boonstra, J. and R.A.L. Kselik, 2002, SATEM: Selected Aquifer Test Evaluation Methods, Publication 48, International Institute for Land Reclamation and Improvement, Wageningen, The Netherlands.

Boulton, N.S., 1954a, The drawdown of the water table under non-steady conditions near a pumped well in an unconfined formation, *Proc. Inst. Civ. Engineers*, 3(3), 564–579.

Boulton, N.S., 1954b, Unsteady Radial Flow to a Pumped Well Allowing for Delayed Yield from Storage, Publication 37, International Association of Scientific Hydrology, pp. 472–477.

Bourdet, D., J.A. Ayoub, and Y.M. Pirard, 1989, Use of pressure derivative in well-test interpretation, *SPE Formation Evaluation,* 4(2), 293–302.

Bouwer, H., 1989, The Bouwer and Rice slug test — an update, *Ground Water*, 27(3), 304–309.

Bouwer, H., and R.C. Rice, 1976, A slug test for determining hydraulic conductivity of unconfined aquifers with completely or partially penetrating wells, *Water Resour. Res.,* 12(3), 423–428.

Bredehoeft, J.D., H.H. Cooper, and I.S. Papadopulos, 1966, Inertial and storage effects in well-aquifer systems: an analog investigation, *Water Resour. Res.,* 2(4), 697.

Burbey, T.J., 1999, Effects of horizontal strain in estimating specific storage and compaction in confined and leaky aquifer systems, *Hydrogeol. J.,* 7, 521–532.

Butler, J.J., 1998a, The Dipole Test for Site Characterization: Some Practical Considerations, Kansas Geological Survey, Open-File Report 98-20.

Butler, J.J., 1998b, *The Design, Performance, and Analysis of Slug Tests,* CRC Press / Lewis Publishers, Boca Raton, FL, p. 252.

Butler, J.J., A.A. Lanier, J.H. Healy, and S.M. Sellwood, 2000, Direct-Push Hydraulic Profiling in an Unconsolidated Alluvial Aquifer, Kansas Geological Survey, Open-File Report 2000-62.

Butler, J.J., C.D. McElwee, and G.C. Bohling, 1999, Pumping tests in networks of multilevel sampling wells: motivation and methodology, *Water Resour. Res.,* 35(11), 3553–3560.

Butler, J.J. et al., 2002, Hydraulic tests with direct-push equipment, *Ground Water,* 40(1), 25–36.

Butler, J.J., Jr., 1991, A stochastic analysis of pumping tests in laterally non-uniform media, *Water Resour. Res.,* 27(9), 2401–2414.

Butler, J.J., Jr., 1988, Pumping tests in non-uniform aquifers — the radially symmetric case, *J. Hydrology,* 101, 15–30.

Butler, J.J., Jr., 1990, The role of pumping tests in site characterization: Some theoretical considerations, *Ground Water,* 28(3), 394–402.

Butler, J.J., Jr., and J.M. Healey, 1988, Relationship between pumping test and slug test parameters: scale effect or artifact? *Ground Water*, 36(2), 305–315.

Butler, J.J. and M.-S. Tsou, 2000, The StrpStrm model (v 1.0) for Calculation of Pumping Induced Drawdown and Stream Depletion, Kansas Geological Survey, Computer Series Report 99-1.

Butler, J.J. and M.-S. Tsou, 2001, Mathematical Derivation of Drawdown and Stream Depletion Produced by Pumping in the Vicinity of Finite-Width Stream of Shallow Penetration, Kansas Geological Survey, Open-File Report 2000-8.

Butler, J.J., V.A. Zlotnik, and M.-S. Tsou, 2001, Drawdown and depletion produced by pumping in the vicinity of a partially penetrating stream, *Ground Water*, 39(5), 651–659.

Butler, J.J. and W. Liu, 1993, Pumping tests in nonuniform aquifers: the radially asymmetric case, *Water Resour. Res.,* 29(2), 259–269.

Butler, J.J. and W. Liu, 1991, Pumping tests in nonuniform aquifers: the linear strip case, *J. Hydrology*, 128, 69–99.

Calver, A., 2001, Riverbed permeabilities: information from pooled data, *Ground Water,* 39(4), 546–553.

Carrera, J., A. Alcolea, A. Medina, J. Hidalgo, and L.J. Slooten, 2005, Inverse problem in hydrogeology, *Hydrogeol. J.,* 13, 206–222.

Chandler, S., P.N. Kapoor, and S.K. Goyal, 1981, Analysis of pumping test data using Marquardt algorithm, *Ground Water*, 19(3), 225–227.

Chen, X.H., 2000, Measurement of streambed hydraulic conductivity and its anisotropy, *Environ. Geol.*, 39, 1317–1324.

Chen, Xunhong and Xi Chen, 2003, Sensitivity analysis and determination of streambed leakance and aquifer hydraulic conductivity, *J. Hydrology*, 284, 270–284.

Cheng, A.H.-D., 2000, *Multilayered Aquifer Systems — Fundamentals and Applications,* Marcel Dekker, Inc., New York, p. 379.

Chiang, W.-H. and W. Kinzelbach, 2003, *3D–Groundwater Modeling with PMWIN,* Springer-Verlag, New York.

Clark, W.E., 1967, Computing the barometric efficiency of a well, *J. Hydraul. Div.*, 93(HY4), 93–98.

Clarke, D.K., 1987, Microcomputer programs for groundwater studies, in *Developments in Water Science,* 30, Elsevier, Amsterdam.

Cooley, R.L., 1992, A Modular Finite-Element Model (MODFE) for Areal and Axisymmetric Ground-Water-Flow Problems, Part 2: Derivation of Finite-Element Equations and Comparisons with Analytical Solutions, U.S. Geological Survey — Techniques of Water-Resources Investigations Report, book 6, chap. A4, p. 108.

Cooley, R.L. and C.M. Case, 1973, Effect of a water table aquitard on drawdown in an underlying pumped aquifer, *Water Resour. Res.*, 9(2), 434–447.

Cooper, H.H. and C.E. Jacob, 1946, A generalized graphical method for evaluating formation constants and summarizing well-field history, *Trans. Am. Geophys. Union,* 27, 526–534.

Cooper, H.H., Jr., J.D. Bredehoeft, and I.S. Papadopulos, 1967, Response of a finite-diameter well to an instantaneous charge of water, *Water Resour. Res.*, 3(1), 263–269.

Crump, K.S., 1976, Numerical inversion of Laplace transforms using a Fourier series approximation, *J. ACM,* 23(1), 89–96.

Das Gupta, A. and S.G. Joshi, 1984, Algorithm for Theis solution, *Ground Water*, 22(2), 199–206.

Davies, B. and B. Martin, 1979, Numerical inversion of the Laplace transform: a survey and comparison of methods, *J. Computational Phys.*, 33, 1–32.

Davis, D.R. and T.C. Rasmussen, 1993, A comparison of linear regression with Clark's method for estimating barometric efficiency of confined aquifers, *Water Resour. Res.*, 29(6), 1849–1854.

Davis, S.N., 1969, Porosity and permeability in natural materials, in *Flow through Porous Media,* R.J.M. DeWiest, Ed., Academic Press, San Diego.

Davis, S.N. and R.J.M. DeWiest, 1966, *Hydrogeology*, John Wiley and Sons, Inc., New York.

Dawson, K.J. and J.D. Istok, 1991, *Aquifer Testing: Design and Analysis of Pumping and Slug Tests,* Lewis Publishers, Inc., New York, 280 pp.

de Lima, V., 1991, Stream-Aquifer Relations and Yield of Stratified-Drift Aquifers in the Nashua River Basin, Massachusetts, U.S. Geological Survey Water-Resources Investigations Report 88-4147, p. 47.

de Marsily, Ghislain, 1986, *Quantitative Hydrogeology,* Academic Press, Inc., San Diego, 440 pp.

Denis, R.E. and L.M. Motz, 1998, Drawdown in coupled aquifers with confining unit storage and ET reduction, *Ground Water*, 36(2), 201–207.

Deutsch, C.V., 2002, *Geostatistical Reservoir Modeling,* Oxford University Press, New York, p. 400.

Deutsch, C.V. and A.G. Journel, 1997, GSLIB: Geostatistical Software Library and User's Guide, 2nd Edition, Oxford University Press, New York.

Doherty, J., 1994, *PEST–Model-Independent Parameter Estimation,* Watermark Computing, Corinda, Australia, p. 122.

Doherty, J., 2003, Groundwater model calibration using pilot points and regularization, *Ground Water,* 41(2), 170–177.

Doherty, J., 2004a, Groundwater utilities, Part A: Overview.

Doherty, J., 2004b, Groundwater utilities, Part B. Program descriptions.

Doherty, J., 2004c, *PEST Model-Independent Parameter Estimation. User Manual,* 5th ed., Watermark Computing, Corinda, Australia.

Domenico, P.A., 1972, *Concepts and Models in Ground Water Hydrology,* McGraw-Hill Book Company, New York, 405 pp.

Dougherty, D.E., 1989, Computing well hydraulics solutions, *Ground Water,* 27(4), 564–569.

Dougherty, D.E. and D.K. Babu, 1984, Flow to a partially penetrating well in a double-porosity reservoir, *Water Resour. Res.,* 20(8), 1116–1122.

Driscoll, F.G., 1986, *Ground Water and Wells,* 2nd ed., Johnson Division, UOP Inc., St. Paul, MN, 1089 pp.

Duffield, G.M., D.R. Buss, and D.E. Stephenson, 1990, Velocity prediction errors related to flow model calibration uncertainty, in *Calibration and Reliability of Ground Water Modelling,* K. Kovar, Ed., Publication 195, International Association of Hydrogeologic Sciences, Wallingford, pp. 397–406.

Duwelius, R.F., 1996, Hydraulic Conductivity of the Streambed, East Branch Grand Calument River, Northern Lake County, Indiana, U.S. Geological Survey, Water Resources Investigations Report 98-4218, p. 37.

Eden, R.N. and C.P. Hazel, 1973, Computer and graphical analysis of variable discharge pumping test of wells, Inst. Engrs., Australia, *Civil Engng. Trans.,* 5–10.

Erskine, A.D., 1991, The effect of tidal fluctuations on a coastal aquifer in the UK, *Ground Water,* 29(4), 556–562.

Fenske, P.R., 1977, Radial flow with discharging-well and observation-well storage, *J. Hydrology,* 32, 87–96.

Fenske, P.R., 1984, Unsteady Drawdown in the Presence of a Linear Discontinuity, in Ground Water Hydraulics, Monograph 9, American Geophysical Union, Washington, D.C.

Ferris, J.G., 1951, Cyclic Fluctuations of Water Levels as a Basis for Determining Aquifer Transmissivity, Publication 33, Vol. 2, International Association of Scientific Hydrology, General Assembly, Brussels.

Ferris, J.G., D.B. Knowles, R.H. Brown, and R.W. Stallman, 1962, Theory of Aquifer Tests: a Summary of Lectures, U.S. Geological Survey, Water-Supply Paper 1536E, 174 pp.

Fetter, C.W., 2000, *Applied Hydrogeology,* Prentice Hall, Inc., Upper Saddle River, NJ.

Fox, G.A., 2003, Improving MODFLOW's RIVER package for unsaturated stream/aquifer flow, in Proceedings of the 23rd Annual Geophysical Union Hydrology Days, J.A. Ramirez, Ed., March 31–April 2, Fort Collins, CO, American Geophysical Union.

Fox, G.A. and D.S. Durnford, 2002, STRMAQ version 1.01, Department of Civil Engineering, Colorado State University.

Freeze, R. and J.A. Cherry, 1979, *Ground Water,* Prentice-Hall, Inc., Upper Saddle River, NJ, 604 pp.

Ghassemi, F.A., A.J. Jakeman, and G.A. Thomas, 1989, Ground water modeling for salinity management: an Australian case study, *Ground Water,* 27(3), 384–392.

Grimestad, G., 1981, Analysis of data from pumping tests in anisotropic aquifers: equations and graphical solutions, *Water Resour. Res.,* 31(4), 933–941.

Guyonnet, D., S. Mishra, and J. McCord, 1993, Evaluating the volume of porous medium investigated during a slug test, *Ground Water,* 31(4), 627–633.

Halford, K.J. and E.L. Kuniansky, 2002, Documentation of Spreadsheets for the Analysis of Aquifer-Test and Slug-Test Data, U.S. Geological Survey Open-File Report 02-197.

Halford, K.J. and R.T. Hanson, 2002, User Guide for the Drawdown-Limited, Multi-Node Well (MNW) Package for the U.S. Geological Survey's Modular Three-Dimensional Finite-Difference Ground-Water Flow Model, Versions MODFLOW-96 and MODFLOW-2000, U.S. Geological Survey Open-File Report 02-293.

Hall, Phil, 1996, *Water Well and Aquifer Test Analysis,* Water Resources Publications, LLC, Highlands Ranch, Colorado, p. 412.

Hanson, R.T. and S.A. Leake, 1999, Documentation of HYDMOD, a Program for Extracting and Processing Time-Series Data from the U.S. Geological Survey's Modular Three-Dimensional Finite-Difference Ground-Water Flow Model, U.S. Geological Survey Open-File Report 98-564, 57 pp.

Hantush, M.S., 1960, Modification of the theory of leaky aquifers, *J. Geophys. Res.,* 65(11), 3713–3725.

Hantush, M.S., 1961, Drawdown around a partially penetrating well, *J. Hydraul. Div., Proc. Am. Soc. Civ. Eng.,* 87(HY4), 83–98.

Hantush, M.S., 1964, Hydraulics of wells, in *Advances in Hydrosciences,* Vol. 1, V.T. Chow, Ed., Academic Press, San Diego, pp. 281–432.

Hantush, M.S., 1966a, Analysis of data from pumping tests in anisotropic aquifers, *J. Geophys. Res.,* 71(2), 421–426.

Hantush, M.S., 1966b, Wells in anisotropic aquifers, *Water Resour. Res.,* 2(2), 273–279.

Hantush, M.S. and C.E. Jacob, 1955, Non-steady radial flow in an infinite leaky aquifer, *Trans. Am. Geophys. Union,* 36(1), 95–100.

Hantush, M.S. and R.G. Thomas, 1966, A method for analysis of a drawdown test in anisotropic aquifers, *Water Resour. Res.,* 2, 281–285.

Harbaugh, A.W., 2002, A Data-Input Program (MFI2K) for the U.S. Geological Survey Modular Ground-Water Model (MODFLOW-200), U.S. Geological Survey Open-File Report 02-41.

Harbaugh, A.W. and M.G. McDonald, 1996a, User's Documentation for MODFLOW-96, an Update to the U.S. Geological Survey Modular Finite-Difference Ground-Water Flow Model, U.S. Geological Survey Open-File Report 96-485, 56 pp.

Harbaugh, A.W. and M.G. McDonald, 1996b, Programmer's documentation for MODFLOW-96, an update to the U.S. Geological Survey modular finite-difference groundwater flow model, U.S. Geological Survey, Open File Report, 96-486, 220.

Hill, M.C., 1998, Methods and Guidelines for Effective Model Calibration. U.S. Geological Survey Water-Resources Investigations Report 98-4005, p. 90.

Hsieh, P.A., 1996, Deformation-induced changes in hydraulic head during ground-water withdrawal, *Ground Water,* 34(6), 1082–1089.

Hsieh, P.A., S.P. Neuman, G.K. Stiles, and E.S. Simpson, 1985, Field determination of three-dimensional hydraulic conductivity tensor of anisotropic media. 2. Methodology and application to fractured rocks, *Water Resour. Res.,* 21(11), 1667–1676.

Hunt, B., 1999, Unsteady stream depletion from groundwater pumping, *Ground Water,* 37(1), 98–104.

Hunt, B., J. Weir, and B. Clausen, 2001, A stream depletion field experiment, *Ground Water,* 39(2), 283–289.

Hvorslev, M.J., 1951, Time Lag and Soil Permeability in Ground Water Observations, U.S. Army Corps of Engineers, Waterways Experiment Station Bulletin 36, Vicksburg, MS, 50 pp.

Hyder, Z., J.J. Butler, Jr., C.D. McElwee, and W.Z. Lui, 1994, Slug in partially penetrating wells, *Water Resour. Res.,* 30(11), 2945–2957.

Indelman, P., A. Fiori, and G. Dagen, 1996, Steady flow towards wells in heterogeneous formations: mean head and equivalent conductivity, *Water Resour. Res.,* 32(7), 1975–1983.

Isaaks, E.H. and R.M. Srivastava,1989, *An Introduction to Applied Geostatistics,* Oxford University Press, New York.

Istok, J., 1989, Groundwater Modeling by the Finite-Element Method, 1st ed., Monograph 13, American Geophysical Union, Washington, D.C., p. 405.

Jacob, C.E., 1940, On the flow of water in an elastic artesian aquifer, *Trans. Am. Geophys. Union,* 21(2), 574–586.

Jacob, C.E., 1944, Notes on determining permeability by pumping tests under water table conditions, U.S. Geological Survey, Mines Report.

Jacob, C.E., 1947, Drawdown test to determine effective radius of artesian aquifer, *Trans. Am. Soc. Civ. Eng.,* 112, paper 2321, 1047–1064.

Johns, R.A., L. Semprini, and P.V. Roberts, 1992, Estimating aquifer properties by nonlinear least-squares analysis of pump test response, *Ground Water*, 30(1), 68–77.

Jorgensen, D.G., 1988, Estimating permeability in water-saturated formations, *The Log Analyst*, 29(6), 401–409.

Kabala, Z.J., 2001, Sensitivity analysis of a pumping test on a well with wellbore storage and skin, *Adv. Water Resour.,* 24, 483–504.

Karasaki, K., J.C.S. Long, and P.A. Witherspoon, 1988, Analytical models of slug tests, *Water Resour. Res.,* 24(1), 115–126.

Kashyap, D.P., P. Dachadesh, and L.S.J. Sinha, 1998, An optimization model for analysis of test pumping data, *Ground Water*, 26(3), 289–297.

Kim, J.-M. and R.R. Parzak, 1997, Numerical simulation of the Noordbergum effect resulting from groundwater pumping in a layered aquifer system, *J. Hydrology,* 202, 231–243.

Kinzelbach, W., 1986, *Ground Water Modeling — An Introduction with Sample Programs in BASIC,* Elsevier Science Publishers, Amsterdam.

Knight, J.H. and G.J. Kluitenberg, 2005, Some analytical solutions for sensitivity of well tests to variations in storativity and transmissivity, *Adv. Water Resour.,* 28, 1057–1075.

Kolm, K.E., 1993, Conceptualization and Characterization of a Hydrologic System, Publication GWMI 93-01, International Ground Water Modeling Center, Golden, CO, 58 pp.

Konikow, L.F., 1978, Calibration of ground-water models, In *Verification of Mathematical and Physical Models in Hydraulic Engineering,* American Society of Civil Engineers, Reston, VA, pp. 87–93.

Kontis, A.L. and R.J. Mandle, 1988, Modifications of a Three-Dimensional Ground-Water Flow Model to Account for Variable Density and Effects of Multiaquifer Wells, U.S. Geological Survey Water-Resources Investigations Report 87-4265.

Kozeny, J., 1933, Theorie und Berechnung der Brunnen Wasserkraft und Wassenwirtschaft. Vol. 28.

Kruseman, G.P. and N.A. de Ridder, 1994, Analysis and Evaluation of Pumping Test Data, 2nd ed., Publication 47, International Institute for Land Reclamation and Improvement. Wageningen, The Netherlands, 377 pp.

Landon, M.K., D.L. Rus, and E.E. Harvey, 2001, Comparison of instream methods for measuring hydraulic conductivity in sandy streambeds, *Ground Water,* 39(6), 870–885.

Leake, S.A., P.P. Leahy, and A.S. Navoy, 1994, Documentation of a Computer Program to Simulate Transient Leakage from Confining Units Using the Modular Finite-Difference Ground-Water Flow Model, U.S. Geological Survey Open-File Report 94-59, 70 pp.

Lebbe, L.L., 1999, *Hydraulic Parameter Identification — Generalized Interpretation Method for Single and Multiple Pumping Tests,* Springer-Verlag Telos, New York, p. 359.

Lee, T.-C., 1999, *Applied Mathematics in Hydrogeology,* CRC Press, Boca Raton, FL, p. 382.

Lennox, D.H., 1966, Analysis of step-drawdown test, *J. Hydraul. Div., Proc. Am. Soc. Civ. Eng.,* 104(HY6), 432–446.

Leven, C., 2002, Effects of Heterogeneous Parameter Distributions on Hydraulic Tests — Analysis and Assessment. Zenrium fur Angewandte Geowissenschafter, Angewandte Geologie, Sigwartstr, 10, 72076 Tubingen, Germany, TGA, C65.2002. Available at w210.ub.uni-tuebingen,de/dbt/volltexte/2003/710/pdf/TGA65-Leven.pdf.

Lohman, S.W., 1972, Ground-Water Hydraulics, U.S. Geological Survey, Professional Paper 708, 70 pp.

Mallet, A., 2000, Numerical Inversion of Laplace Transform. Available at http://library.wolfram.com/infocenter/MathSource/2691/.

Maslia, M.L. and R.B. Randolph, 1987, Methods and Computer Program Documentation for Determining Anisotropic Transmissivity Tensor Components of Two-Dimensional Ground Water Flow, U.S. Geological Survey, Water-Supply Paper 2308, 46 pp.

Maxey, G.B., 1964, Hydrostratigraphic units, *J. Hydrology,* 2, 124–129.

McDonald, M.G. and A.W. Harbaugh, 1988, A Modular Three-Dimensional Finite-Difference Ground-Water Flow Model, U.S. Geological Survey Techniques of Water-Resources Investigations, Book 6, Chap. A-1, 586 pp.

McElwee, C.D., J.J. Butler, Jr., and G.C. Bohling, 1992, Nonlinear Analysis of Slug Tests in Highly Permeable Aquifers Using a Hvorslev–type Approach, Kansas Geological Survey, Open-File Report 92-39.

McKinley, R.M. and T.D. Streltsova, 1993, Monographs for analysis of pressure buildup data influence by heterogeneity, *SPE Formation Evaluation*, Society of Petroleum Engineers.

Millham, N.P. and B.I. Howes, 1995, A comparison of methods to determine *K* in a shallow coastal aquifer, *Ground Water,* 33(1), 49–57.

Moench, A.F., 1984, Double-porosity models for a fissured ground water reservoir with fracture skin, *Water Resour. Res.,* 20, 831–846.

Moench, A.F., 1985, Transient flow to a large-diameter well in an aquifer with storative semiconfining layers, *Water Resour. Res.,* 21(8), 1121–1131.

Moench, A.F., 1994, Specific yield as determined by type-curve analysis of aquifer-test data, *Ground Water,* 32(6), 949–957.

Moench, A.F., 1997, Flow to a well of finite diameter in a homogeneous anisotropic water table aquifer, *Water Resour. Res.,* 33(6), 1397–1407.

Moench, A.F., 1998, Correction to "flow to a well of finite diameter in a homogeneous, anisotropic water table aquifer," *Water Resour. Res.*, 34(9), 2431–2432.

Moench, A.F. and A. Ogata, 1984, Analysis of constant discharge wells by numerical inversion of Laplace transform solutions, in *Ground Water Hydraulics*, J.S. Rosenshein and G.D. Bennett, Eds., Water Resources, Monograph Series 9, American Geophysical Union, Washington, D.C., pp. 46–170.

Moench, A.F. and P.A. Hsieh, 1985, Analysis of Slug Test Data in a Well with Finite Thickness Skin, in International Association of Hydrology Memoirs. Vol. XVII, Part 1, Proceedings of the 17th International Association of Hydrology Congress on the hydrology of rocks of low permeability. Tucson, AZ, pp. 17–29.

Moench, A.F., S.P. Garabedian, and D.R. LeBlanc, 2001, Estimation of Hydraulic Parameters from an Unconfined Aquifer Test Conducted in a Glacial Outwash Deposit, Cape Cod, Massachusetts, U.S. Geological Survey Professional Paper 1629, 69 pp.

Morris, D.A. and A.I. Johnson, 1967, Summary of Hydrologic and Physical Properties of Rock and Soil Materials, as Analyzed by the Hydrologic Laboratory of the U.S. Geological Survey 1948–1960, U.S. Geological Survey, Water-Supply Paper 1839-D.

Mukhopadhyay, A., 1985, Automatic derivation of parameters in a nonleaky confined aquifer with transient flow, *Ground Water*, 23(6), 806–811.

Mukhopadhyay, A., 1988, Automatic computation of parameters for leaky confined aquifers, *Ground Water*, 26(4), 500–506.

Muskat, M., 1937, *The Flow of Homogeneous Fluids through Porous Media,* McGraw-Hill Book Company, New York, 763 pp.

Narasimhan, T.N. and M. Zhu, 1993, Transient flow of water to a well in an unconfined aquifer: applicability of some conceptual models, *Water Resour. Res.,* 29, 179–191.

Neuman, S.P., 1975a, Analysis of pumping test data from anisotropic unconfined aquifers considering delayed gravity response, *Water Resour. Res.,* 11(2), 329–342.

Neuman, S.P., 1975b, A computer program (DELAY2) to calculate drawdown in an anisotropic unconfined aquifer with a partially penetrating well, unpublished manuscript.

Neuman, S.P. and P.A. Witherspoon, 1969a, Applicability of current theories of flow in leaky aquifers, *Water Resour. Res.,* 5(4), 817–829.

Neuman, S.P. and P.A. Witherspoon, 1969b, Theory of flow in a confined two aquifer system, *Water Resour. Res.,* 5(4), 803–816.

Neuman, S.P. and P.A. Witherspoon, 1972, Field determination of the hydraulic properties of leaky multiple aquifer systems, *Water Resour. Res.,* 8(5), 1284–1298.

Neuman, S.P., G.R. Walter, H.W. Bently, J.J. Ward, and D.D. Gonzalez, 1984, Determination of horizontal anisotropy with three wells, *Ground Water,* 22, 66–72.

Nind, T.E.W., 1965, Influences of absolute and partial hydrologic barriers on pump test results, *Can. J. Earth Sci.,* 2, 309–323.

Norris, S.E., 1983a, Aquifer tests and well field performance, Scioto River Valley, near Piketon, Ohio: Part I, *Ground Water,* 21, 287–292.

Norris, S.E., 1983b, Aquifer tests and well field performance, Scioto River Valley, near Piketon, Ohio: Part II, *Ground Water,* 21, 438–444.

Nortz, P.E., E.S. Blair, A. Ward, and D. White, 1994, Interactions between an alluvial aquifer wellfield and the Scioto River, Ohio, *Hydrogeology J.,* 2, 23–24.

Novakowski, K.S., 1989, Analysis of pulse interference tests, *Water Resour. Res.,* 35(11), 2377–2387.

Novakowski, K.S., 1990, Analysis of aquifer tests conducted in fracture rock: a review of the physical background and the design of a computer program for generating type curves, *Ground Water*, 28(1), 99–105.

Nwankor, G.I., J.A. Cherry, and R.W. Gilham, 1984, A comparative study of specific yield determinations for a shallow sand aquifer, *Ground Water*, 22(6), 764–772.

Nwankor, G.I., R.W. Gillham, G. van der Kamp, and F.F. Akindunni, 1992, Unsaturated and saturated flow in response to pumping of unconfined aquifer. Field evidence of delayed yield, *Ground Water*, 30, 690–700.

Oliver, D.S., 1993, The influence of nonuniform transmissivity and storativity on drawdown, *Water Resour. Res.*, 29(1), 169–178.

Osman, Y.Z. and M.P. Bruen, 2002, Modelling stream-aquifer seepage in an alluvial aquifer: an improved lossing-stream package for MODFLOW, *J. Hydrology*, 264, 69–89.

Paillet, F.L., El Sayed Zaghloul, and Tag El Dafter, 1990, Application of Geophysical Well Log Analysis to Characterization of Aquifers in the Sinai Region, Republic of Egypt, U.S. Geological Survey Water-Resources Investigations Report 90-4194, 54 pp.

Papadopulos, I.S., 1965, Nonsteady Flow to a Well in an Infinite Anisotropic Aquifer, paper presented at International Association of Scientific Hydrology Symposium, Dubrovnik, Croatia, October, 1965, pp. 21–31.

Papadopulos, I.S. and H.H. Cooper, Jr., 1967, Drawdown in a well of large diameter, *Water Resour. Res.*, 3(1), 241–244.

Papadopulos, I.S., J.D. Bredehoeft, and H.H. Cooper, Jr., 1973, On the analysis of slug test data, *Water Resour. Res.*, 9(4), 1087–1089.

Peaceman, D.W., 1983, Interpretation of well-block pressures in numerical reservoir simulation with nonsquare grid blocks and anisotropic permeability, *Soc. Pet. Eng. J.*, 23(3), 531–543.

Pedrosa, O.A. and K. Aziz, 1986, Use of a hybrid grid in reservoir simulation, *Soc. Pet. Eng. J.*, 1(6), 611–621.

Peres, A.M.M., M. Onur, and A.C. Reynolds, 1989, A new analysis procedure for determining aquifer properties from slug test data, *Water Resour. Res.*, 25(7), 1591–1602.

Peterson, D.M., 1989, Modeling the effects of variably saturated flow on stream losses, In *Solving Ground Water Problems With Models*, Proceedings of the Fourth International Conference on the Use of Models to Analyze and Find Working Solutions to Ground Water Problems, February 7–9, 1989, National Water Well Associations and International Ground Water Modeling Center, Westerville, OH, pp. 899–928.

Poeter, E.P. and M.C. Hill, 1997, Inverse models: a necessary next step in ground-water modeling, *Ground Water*, 35(2), 250–260.

Poeter, E.P. and M.C. Hill, 1998, Documentation of UCODE, a Computer Code for Universal Inverse Modeling, U.S. Geological Survey Water-Resources Investigations Report 98-4080, 116 pp.

Polubarinova-Kochina, P.Ya., 1962, *Theory of Ground Water Movement,* Princeton University Press, Princeton, NJ, 613 pp.

Prince, K.R. and B.J. Schneider, 1989, Estimation of Hydraulic Characteristics of the Upper Glacial and Magothy Aquifers at East Meadow, New York, by Use of Aquifer Tests, U.S. Geological Survey Water-Resources Investigations Report 87-4211, 43 pp.

Ramey, H.J. and R.G. Agarwal, 1972, Annulus unloading rates as influenced by wellbore storage and skin effect, *Trans. Soc. Pet. Eng.*, 253, 453–462.

Rasmussen, T.C. and L.A. Crawford, 1997, Identifying and removing barometric pressure effects in confined and unconfined aquifers, *Ground Water,* 35(3), 502–511.

Rasmussen, W.C., 1964, Permeability and Storage of heterogeneous Aquifers in the U.S., Publication No. 64, International Association of Scientific Hydrology, pp. 317–325.

Raudkivi, A.J. and R.A. Callander, 1976, *Analysis of Ground Water Flow,* John Wiley & Sons, Inc., New York.

Reed, J.E., 1980, Type Curves for Selected Problems of Flow to Wells in Confined Aquifers, Techniques of Water Resources Investigations of the U.S. Geological Survey, Book 3, Chap. B3, 106 pp.

Reilly, T.E., 1984, A Galerkin Finite Element Flow Model to Predict the Transient Response of a Radially Symmetric Aquifer, U.S. Geological Survey Water-Supply Paper 2198, 33 pp.

Reilly, T.E. and A.W. Harbaugh, 1993a, Simulation of cylindrical flow to a well using the U.S. Geological Survey modular finite-difference ground-water flow model, *Ground Water,* 31(3), 489–494.

Reilly, T.E. and A.W. Harbaugh, 1993b, Source code for the computer program and sample data set for the simulation of cylindrical flow to a well using the U.S. Geological Survey Modular Finite-Difference Ground-Water Flow Model, U.S. Geological Survey Open-File Report 92-659, 7 pp.

Reilly, T.E. and A.W. Harbaugh, 2004, Guidelines for Evaluating Ground-Water Flow Models, U.S. Geological Survey Scientific Investigations Report 2004-5038, 30 pp.

Remson, I., G.M. Hornberger, and F.J. Molz, 1971, *Numerical Methods in Subsurface Hydrology,* Wiley-Interscience, New York, 389 pp.

Renard, P., 2005, The future of hydraulic tests, *J. Hydrogeology,* 13, 259–262.

Rodriques, J.D., 1983, The Noordbergum effect and the characterization of aquitards at the Riop Maion mining project, *Ground Water,* 21(2), 200–207.

Rorabaugh, M.I., 1953, Graphical and theoretical analysis of step-drawdown test of artesian well, *Trans. Am. Soc. Civ. Eng.,* 79, 1–23.

Rorabaugh, M.I., 1956, Ground Water in Northeastern Louisville and Kentucky with Reference to Induced Infiltration, U.S. Geological Survey, Water-Supply Paper 1360-B.

Rushton, K.R. and S.C. Redshaw, 1979, *Seepage and Ground Water Flow,* John Wiley and Sons, Ltd., New York, 339 pp.

Rutledge, A.T., 1991, An Axisymmetric Finite-Difference Flow Model to Simulate Drawdown in and Around a Pumped Well, U.S. Geological Survey. Water-Resources Investigations Report 96-4098.

Saleem, Z.A., 1970, A computer method for pumping-test analysis, *Ground Water,* 8(5), 21–24.

Sanchez-Vila, X., P.M. Meirs, and J. Carrera, 1999, Radially convergent flow in heterogeneous porous media, *Water Resour. Res.*, 33(7), 1633–1641.

Sanchez-Vila, X., 1999, Pumping tests in heterogeneous aquifers: An analytical study of what can be obtained from their interpretation using Jacob's method, *Water Resour. Res.,* 35(4), 943–952.

Schroth, B. and T.N. Narasimhan, 1997, Application of a numerical model in the interpretation of a leaky aquifer test, *Ground Water,* 35(2), 371–375.

Seaber, P.R., 1988, Hydrostratigraphic units, in *Hydrogeology,* W. Bach, J.S. Rosenshein, and P.R. Seaber, Eds., The Geology of North America, Vol. 0-2, Geological Society of America, Boulder, CO, pp. 9–14.

Serfes, M.E., 1991, Determining the mean hydraulic gradient of ground water affected by tidal fluctuations, *Ground Water,* 29(4), 549–555.

Spane, F.A., 1999, Effect of Barometric Fluctuations on Well Water-Level Measurements and Aquifer Test Data, Report PNNL-13078, Pacific Northwest National Laboratory, Richland, WA.

Spane, F.A., 2002, Considering barometric pressure in groundwater flow investigations, *Water Resour. Res.,* 38(6), 14–18.

Spane, F.A., Jr., and S.K. Wurster, 1993, DERIV: A program for calculating pressure derivatives for use in hydraulic test analysis, *Ground Water,* 31(5), 814–822.

Spitz, K. and J. Moreno, 1996, *A Practical Guide to Ground Water and Solute Transport Modeling,* John Wiley & Sons, Inc., New York, pp. 341–354.

Springer, R.K. and L.W. Gelhar, 1991, Characterization of Large-Scale Aquifer Heterogenity in Glacial Outwash by Analysis of Slug Tests with Oscillatory Responses, Cape Cod, Massachusetts, U.S. Geological Survey Water Resources Investigations Report 91-4034, p. 36.

Stallman, R.W., 1971, Aquifer-Test Design, Observation and Data Analysis. U.S. Geological Survey. Techniques of Water-Resources Investigations. Book 3, Chap. B1.

Stehfest, H., 1970a, Algorithm 368 numerical inversion of Laplace transforms, *Commun. ACM,* 13(1), 47–49.

Stehfest, H., 1970b, Remark on algorithm 368 (D5) numerical inversion of Laplace transforms, *Commun. ACM,* 13(10), 624.

Stone, W.J., 1999, *Hydrogeology in Practice, a Guide to Characterizing Ground-Water Systems,* Prentice-Hall, Upper Saddle River, NJ.

Streltsova, T.D., 1988, *Well Testing in Heterogeneous Formations — An Exxon Monograph,* John Wiley and Sons, Inc., New York, 413 pp.

Talbot, A., 1979, The accurate numerical inversion of Laplace transforms, *J. Inst. Math Appl.,* 23, 97–120.

Theis, C.V., 1935, The relation between the lowering of the piezometric surface and the rate and duration of discharge of a well using ground water storage, *Trans. Am. Geophys. Union,* 16, 519–524.

Tongpenyai, Y. and R. Raghaveeen, 1981, The effect of wellbore storage and skin on interference test data, *J. Pet. Technol.,* 33, 151–160.

Torak, L.J., 1993, A Modular Finite-Element Model (MODFE) for Areal and Axisymmetric Ground-Water-Flow Problems, Part 1: Model Description and User's Manual. U.S. Geological Survey — Techniques of Water-Resources Investigations Report, Book 6, Chap. A3, 136 pp.

Trefry, M.G. and C.D. Johnston, 1998, Pumping test analysis for a tidally forced aquifer, *Ground Water,* 36(3), 427–433.

Trescott, P.C., G.F. Pinder, and S.P. Larson, 1976, Finite-Difference Model for Aquifer Simulation in Two Dimensions with Results of Numerical Experiments, U.S. Geological Survey — Techniques of Water-Resources Investigations Report, Book 7, Chap. C1, 116 pp.

Van der Kamp, G., 1976, Determining aquifer transmissivity by means of well response tests: the under-damped case, *Water Resour. Res.,* 12, 71–77.

Van der Kamp, G., 1985, Brief quantitative guidelines for the design and analysis of pumping tests, in Hydrology in the Service of Man. Memoirs of the 18th Congress, International Association of Hydrogeology, Cambridge, pp. 197–206.

Van Rooy, D., 1988, A Note on the Computerized Interpretation of Slug Test Data, Institute Hydrodynamics Hydraulic Engineering Programs, Report 66, Technical University, Denmark, p. 47.

Vasco, D.W., H. Keers, and K. Karasaki, 2000, Estimation of reservoir properties using transient pressure data: an asymptotic approach, *Water Resour. Res.,* 36(12), 3447–3465.

Verruijt, A., 1969, Elastic storage of aquifers, in *Flow through Porous Media,* R.J.M. DeWeist, Ed., Academic Press, San Diego, pp. 331–376.

Walker, D.D. and R.M. Roberts, 2003, Flow dimensions corresponding to hydrogeologic conditions, *Water Resour. Res.,* 39(12), 1349–1357.

Walton, W.C., 1962, Selected Analytical Methods for Well and Aquifer Evaluation, Illinois State Water Survey, Bulletin No. 49, 81 pp.

Walton, W.C., 1963, Estimating the Infiltration Rate of a Streambed by Aquifer-Test Analysis, report to International Association of Scientific Hydrology, General Assembly, Berkeley.

Walton, W.C., 1970, *Groundwater Resource Evaluation*, McGraw-Hill, Inc., New York.

Walton, W.C., 1991, *Principles of Ground Water Engineering*, Lewis Publishers, Inc., New York, 546 pp.

Walton, W.C., 1996, *Designing Ground Water Models with Windows Software,* Lewis Publishing, Inc., New York.

Way, S.C. and C.R. McKee, 1982, In-situ determination of three-dimensional aquifer permeabilities, *Ground Water,* 20(5), 594–603.

Weeks, E.P., 1969, Determining the ratio of horizontal to vertical permeability by aquifer test analysis, *Water Resour. Res.,* 5(1), 196–214.

Weeks, E.P., 1977, Aquifer tests — the state of the art in hydrology, in Proceedings of the Invitational Well-Testing Symposium, Berkeley, California, October 19–21, pp. 14–26.

Weeks, E.P., 1979, Barometric fluctuations in wells tapping deep unconfined aquifers, *Water Resour. Res.,* 15, 1167–1176.

Weight, W.D., 2001, *Manual of Applied Field Hydrogeology,* McGraw Hill, Inc., New York.

Wilson, J.L. and P.J. Miller, 1978, Two-dimensional plume in uniform ground-water flow, *J. Hydraul. Div.,* 104(HY4), 503–514.

Yager, R.M., 1993, Estimation of Hydraulic Conductivity of a Riverbed and Aquifer System on the Sussquehanna River in Broome County, New York, U.S. Geological Survey Water-Supply Paper 2387, p. 49.

Yeh, T.-C. and S. Liu, 2000, Hydraulic tomography: development of a new aquifer test method, *Water Resour. Res.,* 36(8), 2095–2105.

Yeh, W.W., 1986, Review of parameter identification procedures in groundwater hydrology: the inverse problem, *Water Resour. Res.*, 22(2), 1119–1128.

Zhan, X. and J.J. Butler, 2005, Mathematical Derivations of Semianalytical Solutions for Pumping-Induced Drawdown and Stream Depletion in a Leaky Aquifer System. Kansas Geological Survey Open-File Report 2005-10.

Zlotnik, V.A., 1994, Interpretation of slug and packer tests in anisotrophic aquifers, *Water Resour. Res.*, 32(3), 1119–1128.

Zlotnik, V.A. and H. Huang, 1999, Effects of shallow penetration and streambed sediments on aquifer response to stream stage fluctuations (analytical model), *Ground Water,* 37(4), 599–605.

Appendix—Notation

A_n	area underlain by aquifer transmissivity T_n, L^2
b	aquifer thickness, L
b_{cev}	equivalent confining unit single-layer thickness, L
b_e	effective screen length, L
b_{ev}	equivalent aquifer single-layer thickness, L
b_i	individual aquifer layer thickness, L
b_{sc}	effective screen length, L
b'	confining unit thickness, L
b'_i	individual confining unit layer thickness, L
b'_b	average block thickness between fissure zones, L
BE	barometric efficiency
C	well loss constant, T^2L^5
C_d	dimensionless damping parameter
C_v	damping parameter
D	vertical distance from aquifer top to top of pumped well screen, L
D'	vertical distance from aquifer top to top of observation well screen, L
d_s	wellbore skin thickness, L
F_a	approximate value of dimensionless drawdown, L
F'	shape factor defined by Hvorslev
F_{sc}	dimensionless specific capacity factor
H	deviation of slugged well head from static, L
h_{dd}	dimensionless drawdown, L
H_0	initial head slugged well displacement, L
H/H_0	dimensionless normalized slugged well head
K_{eh}	equivalent horizontal hydraulic conductivity, L/T
K_f	fissure horizontal hydraulic conductivity, L/T
K_h	aquifer horizontal hydraulic conductivity, L/T
K_v	aquifer vertical hydraulic conductivity, L/T
$K_0(\ldots)$	modified Bessel function of second kind and order 0
K'	confining bed or block vertical hydraulic conductivity, L/T
K'_s	fissure skin hydraulic conductivity, L/T
K'_b	block vertical hydraulic conductivity, L/T
K'_{ev}	equivalent vertical hydraulic conductivity, L/T
K'_i	individual confining unit layer vertical hydraulic conductivity, L/T
K_s	streambed vertical hydraulic conductivity, L/T
K_s/b_s	streambed leakance, $1/T$

L	vertical distance from aquifer top to bottom of pumped well screen, L
L_c	effective column length, L
L_{cf}	capillary fringe thickness, L
L_s	length of screen or open hole, L
L'	vertical distance from aquifer top to bottom of observation well screen, L
M	number of empirical constants for gradual drainage from unsaturated zone
MAE	mean absolute error of residuals
n	drainable filter pack porosity
N	even number of Stehfest terms (4, 6, 8, etc.)
p	Laplace–transform parameter
Q	constant pumped well discharge rate, L^3/T
Q_{ap}	discharge with pumped wellbore storage, L^3/T
Q_{apo}	discharge with pumped and observation wellbore, L^3/T
Q_{aw}	discharge without pumped and observation wellbore storage, L^3/T
Q_{is}	constant image well strength, Step i discharge rate increment, L^3/T
r	distance from pumped or slugged well, L
r_c	well casing radius, L
r_{cal}	residual drawdown or normalized, L
r_{ce}	effective casing radius allowing for presence of pump, L
r_{co}	radius of observation well in interval where water levels are changing, L
r_0	zero-drawdown intercept, L
r_e	equivalent well block radius, L
r_i	distance between pumped well and image well, L
$rk^{1/2}$	factor is Laplace–domain transform solution for set of aquifer and real well conditions
$r_{iw}k^{1/2}$	factor is Laplace–domain transform solution for set of aquifer and image well conditions
r_{nc}	nominal well screen radius, L
r_o	observation well casing radius, L
r_p	pump pipe radius, L
r_{pp}	distance beyond which well partial penetration impacts are negligible, L
r_r	pumping test domain radius, L
r_{rs}	slug test domain radius, L
r_s	radius of near-wellbore altered transmissivity, L
r_w	effective pumped well radius, L
r_{wo}	effective observation well radius, L
r_{wfp}	outer radius of filter pack, L
RMS	root–mean-squared error of residuals
s	aquifer drawdown, L
S	aquifer storativity
S_a	storativity

s_{ad}	adjusted drawdown, L
S_{av}	areal average storativity
s_b	drawdown calculated by MODFLOW at block node, L
s_c	drawdown in confining unit, L
S_{cu}	confined aquifer storativity or unconfined aquifer-specific yield
S_{cy}	confining unit-specific yield
S_f	fissure-specific storage
s_L	drawdown difference per log cycle of time, L
S_m	measured drawdown, L
S_n	aquifer storativity in area n
s_o	measured or calculated drawdown or head, L
s_p	drawdown at pumped well, L
s_{pp}	drawdown due to effects of well partial penetration, L
s_{sc}	well water level change per unit stream stage change, L
S_{se}	equivalent-specific storativity
S_{si}	individual aquifer layer-specific storativity
S_s	fissure-specific storativity
s_t	total drawdown, L
$s_w(t_p)$	total drawdown, L
s_{wL}	component of drawdown due to well loss, L
S_{wsf}	wellbore skin factor
S_y	aquifer-specific yield
S'_b	block-specific storage
S'_s	block-specific storativity
SE	surface water efficiency
SW	dimensionless pumped wellbore skin parameter
t	elapsed time, T
T	aquifer transmissivity, L^2/T
T_{av}	areal average transmissivity, L^2/T
t_b	basic time lag, T
t_{cs}	time period during which confining unit storativity effects are appreciable, T
t_d	delayed gravity drainage duration, T
T_d	aquifer transmissivity beyond the discontinuity, L^2/T
t_D	first derivative of dimensionless time, t_D
t_0	zero-time intercept, T
T_e	effective transmissivity, L^2/T
t_n	time of nth peak or trough in slugged well data, T
T_n	aquifer transmissivity in area n, L^2/T
T_p	aquifer transmissivity between pumped well and discontinuity, L^2/T
t_s	wellbore storage impact duration, T

T_s	near-wellbore altered transmissivity, L^2/T
t_{sL}	time that must elapse before the straight-line technique can be applied, T
T_{xx}	aquifer transmissivity in x direction, L^2/T
T_{yy}	aquifer transmissivity in y direction, L^2/T
t'	time after pumping stopped, T
TE	tidal efficiency
w	deviation of water level from static level, one-half width of streambed, L
W_d	dimensionless pumped wellbore storage parameter
W_{dp}	dimensionless observation well delayed response parameter
x	distance from streambed center to observation well, L
z	vertical distance from aquifer top to base of confining bed piezometer, L
z_a	distance from aquifer base to slugged well screen base, L
z_b	distance from aquifer base to slugged well screen top, L
z_p	vertical distance above aquifer base to center of piezometer screen, L
z_{pd}	depth below aquifer top to top of pumped well screen, L
z_{pl}	depth below aquifer top to bottom of pumped well screen, L
z_1	vertical distance above base of aquifer to bottom of observation well screen, L
z_2	vertical distance above base of aquifer to top of observation well screen, L
ΔB	atmospheric pressure change expressed in feet of water, L
ΔS	surface water stage or tidal change, L
Δs	drawdown per logarithmic cycle, L
Δs_i	increment of drawdown during Period i due to increment of discharge ΔQ_i, L
ΔW	water level change in well, L
Δx	pumped well block grid spacing in x direction, L
Δy	pumped well block grid spacing in y direction, L
β	bulk modulus of compression of groundwater
γ	specific weight of groundwater
θ	aquifer porosity
ω	frequency parameter

Index

D

E

R

S

V

W